AN ECONOMETRIC AND GAME THEORETIC MODEL OF COMMON POOL RESOURCE MANAGEMENT: PEOPLE'S PARTICIPATION IN FOREST MANAGEMENT IN INDIA

An Econometric and Game Theoretic Model of Common Pool Resource Management: People's Participation in Forest Management in India

Wietze Lise

Nova Science Publishers, Inc.
New York

For permission to use material from this book please contact us:
Telephone 631-231-7269; Fax 631-231-8175
Web Site: http://www.novapublishers.com

NOTICE TO THE READER

Library of Congress Cataloging-in-Publication Data
Available upon request

ISBN 13 978-1-59454-397-5
ISBN 10 1-59454-397-6

Published by Nova Science Publishers, Inc. ✣ New York

To Banu, Annem & Heit

In memory of Babam & Mem

Contents

List of Tables

List of Figures

Acknowledgements

The question 'Five years of research in India?' has been asked me on several occasions. In the answer to that question lies my intrinsic motivation for writing this book. In 1988, I became very enthusiastic about an elective course named 'rich and poor countries' which I attended as part of my MSc study of Applied Mathematics at Twente University (the Netherlands). While I was wondering how a study of this subject could seriously be combined with mathematics, I came in contact with Arun BAGCHİ, Professor of Applied Mathematics, Twente University. He came with the suggestion that by undertaking a study in and on India, I could 'experience poverty without sacrificing my career'. This convinced me and I ended up writing my practical training report at the Planning Commission (New Delhi), my master thesis at the Institute of Economic Growth, Delhi (to obtain my MSc degree in 1992); while I completed my PhD thesis at the Delhi School of Economics in August 1997.

Many people have contributed to my work in various ways. Firstly, I am very grateful for the continued support and stimulating ideas of my former PhD supervisors. Gopal K. KADEKODİ, Director Institute for Social and Economic Change, already guided me in writing my master thesis, and has played an important role in guiding my research in 1993-1997. Kaushik BASU, Professor at Delhi School of Economics at that time, has guided my research in 1993-1994, but as he accepted his current position as Professor of Economics at Cornell University, T. C. A. ANANT, Professor at Delhi School of Economics, accepted to guide me in 1994-1997. Secondly, During my field work, people from mediating organisations facilitated me in contacting the right villages, and discussions and interviews with villagers themselves led to many additional insights. I am very appreciative to their help, hospitality, and sharing of their views. Finally, I am also grateful to all other people that contributed to my research.

Between 1999–2004 I have been an environmental economic researcher at the Institute for Environmental Studies, Free University, Amsterdam (the Netherlands). While working in various research projects, I managed to publish three articles on the issue of 'people's participation in forest management in India' in international refereed journals. I thank the publishers for granting me permission to use this material in this book. Firstly, with kind permission from Elsevier, I included [115], © 2000 Elsevier Science B. V. Printed in the Netherlands (used in Chapter 3 and Sections 1.2., 2.1. and 3.3.). Secondly, with kind permission of Springer Verlag and Business Media, I included [116], © 2001 Kluwer Academic Publishers. Printed in the Netherlands (used in Chapter 4). Thirdly, with kind permission of Cambridge University Press I included [119], © 2004 Cambridge University Press. Printed in the United Kingdom (used in Chapter 5 and Sections 2., 2. and 3.). In addition, I am thankful to the Society for Environmental Communications, New Delhi, India for permitting me to include [114], © 1997 Society for Environmental Communications. Printed at Thomson Press, India. (used in Section 3.3.1.).

To make this book attractive to a larger audience, and to give evidence of my linguistic knowledge, I added an overall abstract and got it translated into Turkish and Hindi. I thank Ranjit Singh for the Hindi translation. Special thanks go to my wife Banu BAYRAMO ĞLU LİSE for helping me with the Turkish translation and proof-reading the near-final version of this book.

I am grateful to Nova Science Publishers Inc. for publishing this monograph.

W. L.

Augustinuspark, Amstelveen
October 2004

Abstract

A major dilemma of rural development is to simultaneously preserve forests and combat poverty. Many problems of rural development have emerged by considering the trade-off between forest preservation and poverty reduction as a zero-sum game. Local village level institutions are key to solving this dilemma, by considering people and forests mutually dependent. Thereupon, at the heart of this Book lies, simultaneously, a review of the literature on institutions and game theory, formulation of a repeated game model among people who want to gain from the forest, derivation of an estimation procedure for games, in conjunction with development of a quantitative questionnaire.

The key-features of this Book are, first, an econometrical and game theoretical analysis of the primary data collected by the author. Second, a procedure has been developed for estimating a game, which can also be applied to various other fields and disciplines. Third, a great amount of empirical details is collected about the local situation in three case studies, which are compared with each other to find a general strategy for people's managed forest regeneration.

From a methodological point of view this Book provides a manual for undertaking research into the role of people in commonly shared resources, like forests, water bodies, fisheries and grazing pastures. The method in this Book constructs indicators for the level of participation, net benefits from participating, forest dependence, forest quality, inequality, wealth and social differences. From an Indian policy point of view it can be concluded that (1) the studied organisations for forest management vary considerably in their effectiveness and (2) voluntary participation is most likely to emerge among the villagers who depend highly on the forest and perceive the quality of the forest as good.

संक्षेप

एक मुख्य ग्रामीण विकास की दुविधा है समकालिन जंगलों की सुरक्षित रखना व गरीबी का सामना करना। अनेक समस्यायें देहाती विकास पर प्रकट होती है जब ध्यान में रखते हुए जंगल की विनिमय परस्पर व गरीबी की एक निम्नतम तक गटाना योजना। स्थानीय समानय देहाती संस्थावों अत्यावश्यक है इस दुविधा को हल करने में ध्यान में रखते हुए की जनता और जंगल पारस्पर आश्रित होता। फलस्वरुप इस पुस्तक की गंभिरता इस में है, की समकालिन, एक सर्वेक्षण संस्थावों की साहित्य व योजना की सिद्धान, एक दोहराने योग्य नमुना की प्रतिपादन जो जंगल से लाभ चाहता है, योजना के लिये व्युत्पत्ति अनुमानित कार्य-पद्धति, संयुक्तता के साथ संख्या प्रश्नावली की विकास।

इस पुस्तक का मुख्य विशेषतायें है, पहली, मुख्य जानकारी मितव्ययी व सैद्धांतिक विश्लेषण लेखक द्वारा एकत्रित की गई। दूसरा, कार्यविधी विकसित करके एक अनुमान पर आधारित योजना के लिये, जो अनेक भिन्न क्षेत्रों व अनुशासासनों पर भी लाघू हो सकता है। तीसरा, एक महत्वपूर्ण राशी ब्यौरा अनुभव एवं प्रयोग पर आधरित ज्ञान एकत्रित करके स्थानीय स्थिति के बारे में तीन विषय पर अध्ययन द्वारा, जो आपस में तुलनात्मक हो, एक आम योजना पाने में जनता के लिये पुनरुज्जीवित करी गयी जंगल को।

सुव्यवस्थिती के दृष्टि-कोन के अधारित पर इस पुस्तक एक नियम-पुस्तिका शोध या अनुसंधान करने के लिये जनता के कर्त्तव्यों पर सामानय हिस्सेदारी के संपत्तियों में, जैसे जंगल, मछली-तल-छट, जल-भंढार और गौचर-भुमि। इस पुस्तक का तरीका सूचकों को निमणित करता है भागीदारी की बराबर, शेस लाभ भागीदारी द्वारा, जंगल निर्भरता, जंगल की विशेषता, असमानता, संपत्ति और सामाजिक भिन्नता। भारतिय राजनीतिक दृष्टि-कोन से यह निष्कर्ष निकलता है की (1) अध्ययन की गई संस्थावों जंगलों की प्रबन्ध संचालन के लिये परिवर्तित अपनी प्रभावकारि में महत्त्वपूर्ण और (2) स्वैच्छिक भागीदारी अधिकतम संभव प्रकट होता है देहातीयों के बीच में जो अधिक निर्भर जंगलों पर है और जंगल की विशेषता उत्तम समझता है।

Özet

Kırsal alandaki gelişme ile ilgili en büyük ikilemlerden biri bir yandan ormanları korurken eş zamanlı olarak da yoksullukla mücadelenin gerçekleştirilebilmesidir. Kırsal gelişmeyle ilgili olan problemlerin pek çoğu orman korumacılığı ve yoksullukla mücadele arasında bir sıfır-yekün oyunu olduğu görüşünden ortaya çıkmaktadır. Köy enstitüleri, insanları ve ormanları karşılıklı olarak birbirine bağımlı varlıklar olarak kabul etmek suretiyle bu ikilemi çözmekte kilit bir rol üstlenebilir. Bütün bunlara bağlı olarak bu kitabın özünü organizasyonlar ve oyun teorisi literatürünün gözden geçirilmesi, ormanlardan kazanç sağlamak isteyen kişiler arasında tekrarlanan oyun modelinin biçimlendirilmesi, ve kantitatif bir anket oluşturmakla bağlantılı olmak suretiyle oyunlar için bir tahmin yönteminin türetilmesi oluşturmaktadır.

Bu kitabın temel noktaları şunlardır: ilk olarak, yazar tarafından toplanmış olan verilerin ekonometri ve oyun teorisi açısından bir analizi yapılmıştır. İkinci olarak, diğer başka disiplinlerde de kullanımı mümkün olan bir oyun tahmin yöntemi geliştirilmiştir. Üçüncü olarak da kişilerin katılımıyla oluşturulan düzenli orman yenilenmesi için genel bir strateji bulmak amacıyla bölgesel durumu ortaya koyan ve üç bölgeden toplanmış olan bol miktardaki ampirik veri birbirleriyle kıyaslanmıştır.

Metodolojik açıdan bakıldığı zaman bu eser insanların ormanlar, sular, balıkçılık alanları ve yeşil alanlar gibi ortak olarak paylaşılan ve kullanılan doğal kaynaklar üzerindeki rollerini araştırmak için bir el kitabı sunmaktadır. Bu kitaptaki metod ile katılım seviyesi, katılımdan elde edilen net fayda, orman kalitesi, eşitsizlik, zenginlik ve sosyal farklılıklar açısından göstergeler oluşturulmuştur. Hindistan politikası açısından bakıldığında şu sonuçlar çıkarılabilir: (1) orman idaresi açısından incelenen organizasyonlar etkilikleri açısından çeşitlilikler göstermektedirler, (2) gönüllü katılım en çok ormanlara büyük ölçüde bağımlı olan ve köylerindeki ormanların kalitesini yüksek bulan köylüler arasında oluşmaktadır.

Samenvatting

Een groot dilemma van rurale ontwikkeling is om gelijktijdig bossen te beschermen en armoede te bestrijden. Vele problemen van rurale ontwikkeling zijn ontstaan doordat men de afweging tussen bosbeheer en armoede reductie heeft beschouwd als een nul-somspel. Locale instituties op dorpsnivo vormen een sleutelrol in het oplossen van dit dilemma, waarbij de lokale bevolking en de bossen als wederzijds afhankelijk worden beschouwd. De kern van dit boek bestaat tegelijkertijd uit, een behandeling van de literatuur over instituties en speltheorie, de formulering van een herhaald spel model tussen mensen die baat hebben van de bossen, de afleiding van een schattingsprocedure voor spellen, tezamen met de ontwikkeling van een kwantitatieve enquête.

De belangrijkste onderdelen van dit boek zijn, ten eerste, een econometrische en spel theoretische analyse van de primaire data die zijn verzameld door de schrijver zelf. Ten tweede, een procedure is ontwikkeld voor het schatten van een spel die ook gebruikt kan worden voor andere toepassingen en vakgebieden. Ten derde, een grote hoeveelheid empirische details is verzameld over de lokale situatie in drie verschillende Indiase deelstaten. Zij zijn met elkaar vergeleken om tot een algemene strategie te komen voor een door mensen beheerde bosverbetering.

Vanuit een methodologisch standpunt vormt dit boek een handboek voor het uitvoeren van onderzoek naar de rol van mensen in gezamenlijk beheerde natuurlijke rijkdommen, zoals bossen, meren, visserij gebieden en graas gebieden. Met de methode van dit boek zijn indicatoren ontwikkeld voor de mate van participatie, de netto baten van participatie, bosafhankelijkheid, de kwaliteit van het bos, ongelijkheid, bezit en sociale verschillen. Vanuit een Indiaas beleidsoogpunt kan worden gekonkludeerd dat (1) de bestudeerde organisaties voor bosbeheer nogal verschillen in de mate waarin zij effectief zijn en (2) vrijwillige participatie zich het meest waarschijnlijk zal voordoen tussen dorpelingen die sterk afhankelijk zijn van bossen en die de kwaliteit van hun bossen als goed ervaren.

Part I

LITERATURE & MOTIVATION

Chapter 1

Introduction

1. Purpose and Motivation

Management of land is fundamental to rural development. Among the land resources, forests have played an important role in the Indian rural economy, by linking food, fodder, and fuel as the three basic needs for making a living. It is the forest that provides these basic needs. Therefore, a study of forest management is an integral part for a rural development strategy in India. In addition, the current debate on human induced climate change places the regeneration and preservation of the forests higher on the policy agenda. [1]

1.1. Historical Perspective

The need for people's participation in forest management in India has emerged gradually over time. In the precolonial period, the forests produced more than the people required, because the population was low in number. Except for a few 'royal trees', there were no restrictions on cutting trees.

After the arrival of the British in India, large scale felling of trees was initiated, to secure a good supply of timber to construct ships and, later on, railway sleepers. They considered the forests as inexhaustible. They even granted local people the legal rights to clear forested land from trees, to employ it for agricultural purposes. This led to a rapid decline in the forest cover of India and it was recognised that forests were indeed exhaustible.

To decelerate the decline in forest cover, Dr Dietrich Brandis, a German botanist, was appointed as the *first* inspector general of forests in 1864. In that period, Germany was seen as the leading European nation in forest management (Guha, [72, page 88]). This was the first attempt in forest conservation in India. Contrary to the general belief, Dr Brandis favoured community oriented management of forests near the villages, while more distant forest could be managed by the State [72]. The British leaders, however, decided in favour of maintenance of State control in their forest policy.

The first Indian forest Act was approved in 1865. This Act, which was revised several times, made a classification of forests into reserved and protected forests. While the people lost almost all their rights in reserved forests, their use-rights were maintained in protected

[1]More information on the issue of agenda setting with respect to climate change can, for instance, be found in Bayramoğlu [16] and Bayramoğlu and Lise [17] .

forests. However, local communities were not entrusted in managing the protected forest, because the consensus among the British was that local people could not be held accountable for managing their forests. The British preferred to maintain their monopoly position in forest management. Well-planned plantations through *scientific forestry* were undertaken to secure a steady supply of timber.

The reduction of people's rights in the forests led to great hardship in collecting their resources. Having reduced the access, the same protected forest came into use by more than one community. This led to new conflicts. Community management of forests was not possible anymore. People were alienated from their forests. This led to large scale protests through forest fires all over India. This showed the need for people's participation in forest management and the difficulties in attaining it.

1.2. The Present Situation of Forest Management in India

Nowadays, the Indian forest cover is reducing at an alarming speed. Chakraborty [30] goes as far to say that India may lose all its forests by AD 2010. This calls for serious action to reverse the process of deforestation.

According to Chakraborty [30], there are two diametrically opposite viewpoints to the cause of deforestation in India. On the one hand, the official viewpoint says that an increased demand for fuelwood, timber, land for agricultural expansion and housing space leads to forest depletion. They identify the growth in population and livestock in combination with forest dependence and poverty as the main cause. This official viewpoint states that free rights of the people in the forest has led to depletion, because the people did not contribute much to the maintenance and regeneration of the forests.

On the other hand, the popular perception is that the main cause for deforestation lies in the failure of the forest bureaucracy to secure people's participation [30, page 232]. The forest bureaucracy is characterised by centralised efforts, predetermined outcomes, and it is revenue oriented.

1.3. The Need for People's Participation in Forest Management in India

In a country like India, it is not possible to close the forest for use of the State alone because many people depend on the forest for basic needs like fuelwood, timber, fodder, medicines, etc. Unless people's dependence on the forest is reduced, by giving them better opportunities, it is very unlikely that forests can be managed as a state property regime. In the present situation, the only path to sustainable forest management is by seeking people's participation.

Rather, institutions should emerge at the village level to protect and regenerate the forest. This requires collective action of all involved parties: the State, non-governmental organisations, and the people. The main focus of this Book is on people, to see to what extent people's participation has been secured so far in existing afforestation activities. I make a comparison between three diverse institutional set-ups at the village level which provide incentives for people's participation in reforestation. The hypothesis under question is to find factors that influence people's participation in forest management where participation consists of three components, namely, contribution to, benefiting from and involvement in decision making and evaluation.

Figure 1.1. Map of India and surrounding countries.

2. Selection of Field Sites and the Database

The three institutional settings of Chapters 7–9 have been selected with great care, because of their diversities, to study different strategies in forest preservation. Figure 1.1 provides a map of India, while the geographical location of the field sites is presented in Figure 1.2. Table 1.1 points out the diversities between the three field settings.

2.1. Forest Management in Three Indian States

I conducted a study in three States of India: Haryana, Jharkhand and Uttaranchal in 1995 and 1996, spending a net time-period of four months in the field. Each case attempts to involve local people in forest management. Forest use is coordinated through regular meetings at the village level. Below a summary is given of the situations in the three States of India, while a more extensive presentation, discussion and analysis can be found in Part III of this Book.

In Haryana *joint forest management* (JFM), as the process is named after a resolution by the Indian government in 1993, originated from a siltation problem in the Sukhna lake. Deforestation was identified as the main reason for this siltation problem and the forest department initiated new plantations, but the survival rate of the trees was very low. Representatives of the State went to a village called Sukhomajri, which is situated at the root of the river leading to the Sukhna lake, to inquire about their reason for destroying the new plantations (Mishra [138]; Mishra and Sarin [139]). Daulat Ram, a villager from Sukhomajri, suggested constructing a dam for irrigation water. This negotiation process between

Figure 1.2. Location of the three case study regions in India.

Table 1.1. Basic features of the three case studies on forest management

State	Haryana	Jharkhand	Uttaranchal
Name of local organisation	Hill Resource Management Society	*Gram sabha* (village council)	*Van panchayat* (forest council)
Frequency of the meetings	Monthly	Fortnightly	Quarterly
Started in	1977	1984	1931
Initiated by	State	Non-governmental organisation	People
Property regime	State	Private (group)	Common
Number of organisations[a]	48	44	4645
Forest quality	Degraded	Very degraded	Well-stocked
Kind of forest resources	Grass, scrubs, small trees	Multi-tiered cropping pattern	Pine, oak
Kind of village council	Elected village body	Like Haryana, but whole village is involved	Nested structure

[a]: Situation in 1996.

Source: Lise [112] and reinterpretation of my fieldwork.

the State and villagers resulted in the birth of social fencing where the people protect the new plantations while getting water from the dams. This experience resulted in a strategy for forest officials to conserve and recreate forests by helping the people in return for their participation. Since 1977, the State of Haryana has leased the forests to a number of villages. Villagers had the opportunity to create a council to manage their forest. All residents of the village became members of that council. In most villages the State has built dams to serve a double purpose: to check soil erosion of the hilly area of the forests and to provide irrigation water to the land of the villagers. Besides damwater the villagers can collect resources, such as fodder, fuelwood and fibre grasses (*bhabbar*) from the forest.

Chakriya vikas (rotational development) in Jharkhand is a continuation of the successes in Haryana. It started as a sharing model, under which people could contribute land and labour to resolve a deadlock situation of economic stagnation. The investment strategy is to bank on trees as a long-term sustainable income, while maintaining a short-term crop to meet the direct food demand. A non-governmental organisation encourages villagers to pool private land to be subjected to a multi-tiered cropping pattern including (fruit) trees. The output from the pool is shared in three equal parts among people who pool land, people who plant and maintain the saplings, and a village fund. This is called the '1:1:1' sharing mechanism. This case is very exceptional because villagers in one of the most remote places in India must save 33% of their communal income from the pool. Private production and income are excluded, so the total overall saving rate at the household level is lower, but still remarkable. The sharing mechanism is possible because it is a prerequisite for initiating the process at the village level; it is a pre-negotiated sharing rule from which the process starts. The village fund is to be used for development of the village like building schools, wells, roads, or to maintain the pool. The role of the State is negligible in these projects. The people who work on the pool receive a stipend besides a part of the output from the pool. The pool is generally created on wasteland. The village societies are organised very well, all villagers are involved and meetings are held fortnightly.

*Van panchayat*s (forest councils) in Uttaranchal originated before independence from British rule. Large-scale protests of the people by setting the forest on fire prompted the State to allow the people self-organisation by giving them formal rights for managing a communal forest. It led to an organised way of managing a common-pool resource with a common property regime. An autonomous village organisation emerged to manage a communal forest. Since 1931, the State allows for the creation of forest councils by villagers. A forest council consists of a committee that decides on how to use the forest. Forest councils are organised quite well, perhaps because of a fairly equal land and cattle holding. By their permission, resources such as fuelwood, fodder and timber can be collected from the forest. The role of the State is much less pronounced in Uttaranchal than in Haryana due to the remoteness of the forest which forms a physical barrier, making State control fairly difficult. *Van panchayat*s have a clearly privileged position and that is perhaps the reason why they are against conversion into joint forest management, which is coming up separately in the plains; *van panchayat*s still persist in the hills (Saxena [178]).

Table 1.1 points out the diversities between the three institutional settings. The different amount of State involvement is the most distinguishing factor among the three institutional settings. The rights of the people also differ across the common forests. It is state property in Haryana, common property in Uttaranchal and pooled private property in Jharkhand.

The institutional background is very different in the three cases; for example, each has a different history and different sharing rules for the resources. The institutional settings also differ in forest quality, in types of forest resources and in organisation of the village council. In spite of all these (extreme) differences, there are also things in common. For instance in all three cases forests are managed by a well-defined group of people and people are free to choose their level of participation. Participation of people is the joining element. Therefore, these three case studies give a quite diverse insight into the process of people's participation.

2.2. Sample Selection

The field settings as described in Section 2.1. were selected with great care. This was also true for the villages and households. I based the selection of the main villages on discussions with experienced researchers and field workers. Sometimes I also followed the suggestion of managers from the village committees to visit neighboring villages. In this way, I was able to cover the most important villages, as well as some villages which have never been studied before. An important principle was to capture the diversity among various villages with the same institutional set-up. Moreover, I tried to incorporate the following dichotomies:

- caste homogeneity — caste heterogeneity;
- rich — poor households;
- large — small — landless farmers;
- small — large villages;
- easily accessible — remote villages;
- well-managed — badly-managed forests.

Within the village, I always first oriented myself of the local situation and tried to interview the president of the village committee. The (elected) president is generally quite knowledgeable about the history and the situation in the village, such that a wealth of qualitative information at the village level could be obtained.[2] Following the suggestions of the president, other villagers could be approached as well. In addition to the president, the secretary, treasurer, and a number of managing members of the village committee would be interviewed as well. Since every villager is per definition a member of the local organisation, in the sense that they can benefit from collected products from the forest, they constitute the majority of the interviewees, namely 54% in Haryana, 41% in Jharkhand, and 66% in Uttaranchal (see Table B.6). The first row of Tables B.9, B.13 & B.17 contains the total population in each village, while the first row of Table B.5 contains the average total population in each village. On average, about 15% of the total number of families in the village is covered.

The data have been collected through interviews with representatives of a household by developing a quantitative questionnaire. One purpose of the questionnaire was to figure out to what extent people approved of their village council. It was attempted to interview the respondents in isolation to secure their sincerity in their responses. However, this was

[2]See, for instance, the village level questionnaires in Sections 2., 4. and 6..

not always possible. Their responses were cross-checked by, for instance, comparing their income with their expenditures. This direct feedback on the respondents is one of the advantages of interviews and made the results more reliable.

2.3. Notes on the Field Survey and the Database

In a country like India, 'participatory development' plays a crucial role in combatting poverty.[3] Poverty means that the people are deprived from the opportunity to access basic necessities like health care, education, and housing, in addition to food, fodder and fuel. This can not be provided to all from 'above' through the Indian government. In addition to this 'delivery system by the State' it is also important to empower the people to a certain extent in decision making and management. Burkey [28], for instance, provides a practical framework for improving the ability of the people to make their own decisions. Furthermore, he makes an argument that field research can be successful only if the people can understand why it is being done. This understanding has a stimulating effect on them, and they are more likely to participate and be ready to provide information, as good as they possibly can. In that circumstance it benefits both the researcher and the people under study. The researcher needs information to understand and quantify their problems. Ultimately, the solution should be handed over to them as well, since they have many problems which they need to solve by themselves. Letting the people fully participate in the research probably leads to the most meaningful results.

Burkey also warns against research based on a quantitative analysis alone. While quantifying the problem may be helpful and necessary for a successful analysis of the problem, many other factors can have great importance as well. The people know by themselves what their problems are. They should be given the opportunity to tell their own story, instead of relying on quantitative questions alone.

This is why I followed both a quantitative and a qualitative approach during the field survey. The double approach of the fieldwork by gathering quantitative and qualitative information enables the researcher to cross-check the outcomes of both information sets. After several test-rounds in Haryana, I developed questionnaires (see appendix A) to measure a variety of quantitative and a qualitative information on three levels:

1. The individual level,
2. The household level, and
3. The village level.

Information at the household level is most important, because it provides the necessary information for estimating their role and strategies within their specific institutional context. This information can be used to test the game theoretic model, on which Part II of this Book elaborates. Information at the individual level can be used to derive aggregated indicators at the household level, which can serve as socio-economic variables to be included in a regresssion model, but it can also provide insight into the variation within the sample. Qualitative information at the village level is useful to compare the estimated games with

[3]For a further discussion on the issue of participatory development in India see Kadekodi and Lise [90] and Lise [109, 110], while Lise [108] has a focus on planning in India.

the expected outcome, for verifying the robustness and appropriateness of the model and for interpreting the result.

In total 385 households from 32 villages have been interviewed. The variation in the data sample is presented in Table B.5. Table B.5 shows that the field survey covered 9, 12 and 11 villages and 127, 123 and 135 households in Haryana, Jharkhand and Uttaranchal. Furthermore, Table B.5 splits the number of respondents up by gender, age and education. The Table shows, for instance, that the interviewed households in Uttaranchal are relatively older and more educated than the interviewed households in Haryana and Jharkhand.

The field survey yielded a broad range of quantitative information around the following groups of variables:

- Their perception of the quality of the forest;[4]
- Their level of participation in forest management;[5]
- Family ownership of capital goods, like land, cattle and private assets;
- Family members' income from different sources;
- Family expenditures for food items and clothes;
- Various other indicators, like caste, religion, gender, education, family size, etc.[6]

While the main purpose of the field survey has been to collect data for estimating the game model presented in Part II, the data is also used for finding factors that explain participation by several socio-economic variables. Explanatory variables are: forest quality, forest dependence, age, education, gender, family size, caste, religion, wealth, inequality. Some indicators at the village level are: number of members in the local organisation, size of the forest, village fund, and years of forest protection. The village level questionnaire was more or less open, to generate information about the specific circumstance of the local situation.

The field survey was not solely a method for gathering answers on a predetermined set of questions; it was a learning process to improve the questions and understand which questions could be asked, and which questions have to be omitted. It also provided an understanding for interpreting the results. The ultimately used questionnaires can be better seen as the outcome of the research, rather than the starting point. Especially during the initial test-phase of the questionnaire, some villages were visited repeatedly. But as soon as the questionnaire was satisfactory, most villages were visited only once, sometimes twice.

[4]Poteete and Ostrom [164], for instance, argue that asking for the perception of the quality of the forest may be more useful than physically measuring the quality of the forest, as 'local action depends on perceived benefits from the forest and threats to them' [164, page 22].

[5]An important issue is whether people had an incentive to lie during the surveys. This incentive was there during the test phase of the interviews, as they thought the interviewer to represent the government. But by starting the interviews with psychological response questions concerning their level of participation, this suspicion was eliminated.

[6]The detailed questionnaires, which has been adapted to the State specific situation, can be found in Appendix A.

3. Outline of the Book

The main research questions of this Book are the following.

- Which factors enhance people's participation in forest management?
- How can a game model be estimated from primary survey data?
- What can be suggested by application of a (repeated) game model of people's participation in forest management?
- What can we learn from institutions that enhance people's participation in forest management?

The second question is mainly methodologically oriented, while the first and last two questions are directed towards policy making with respect to forest management. To address these research questions, the Book is divided into the following four parts:

I. A survey of the literature on institutional economics, the general problem of managing a common pool resource (CPR), and game theoretic approaches to this problem.
II. An econometric analysis to find factors which explain people's participation in forest management, while an estimation method is derived for and applied to a (repeated) game model of people's participation in forest management, to study incentives for voluntary participation.
III. A description and analysis of the institutional setting in the field, where anomalies in particular villages are pointed out as well.
IV. A statistical Appendix with the questionnaires used to derive the data, and a presentation of key indicators at the State and the village level.

The outline of this Book is as follows. Chapter 2, is basically a survey of the literature. I discuss the differences between institutions and organisations, while the main focus is on institutions. Different property regimes are linked to degrees of 'ownership.' The literature survey is illustrative in the sense that the problem of forest management belongs to the class of common pool resource problems, which require an institutional economic approach. I give an overview of game theoretic approaches to institutions in general and common pool resource problems in specific, to motivate the choice of the model in Part II of this Book.

In Part II, Chapter 3 analyses the data set as derived through my field survey with two techniques. A factor analysis is performed on 10 to 12 participatory indicators of each household. A regression analysis is carried out using the primary data.

In Chapter 4, I develop two methods to estimate the payoff matrix. I give a taxonomy of (repeated) two-person symmetric games. The possibility to measure the choice and the added value to that choice for a person through a primary survey is suggested. The basic idea is to split the interviewees into four groups, where each group represents one of the four cells of a two person symmetric binary choice payoff matrix. In an example, I illustrate the developed estimation techniques. I discuss possibilities for extending the model to more than two players, multiple strategies and asymmetric games.

Chapter 5 applies the general model as developed in Chapter 4 to the situation of people's participation in forest management. For the taxonomy of games in Chapter 4, trigger strategies are developed to derive conditions under which mutual participation can be sustained in equilibrium. Games are estimated at the State and village level and compared to the actual situation in the field.

A final Chapter is added to Part II. Chapter 6 extends the game of the previous Chapter by considering trigger strategies with a finite number of deviations. One scenario, where the payoff decreases due to resource depletion under no participation, is worked out for the case with changing payoffs to show how the stage game could differ over time.

In Part III, I discuss the three rural settings. The structure of each case study is quite similar. Each institutional setting forms a Chapter. These Chapters are divided into several Sections. I give a short introduction and motivation for each Chapter in the first Section. The second Section is devoted to a short historical background of the process, where I explain the general structure of the process with respect to people's participation, benefit sharing and involvement of other actors. In a third Section, I report anomalies from the general pattern and the actual experience in the field in a few selected case studies at the village level. Based on expectations and actual experiences, I present the shortcomings and practical limitations in implementing the process in the fourth Section. In the final Section, I provide suggestions for improving the process.

In Chapter 10, I summarise the main results, draw overall conclusions and point out areas for further research.

Part IV is the Appendix. Appendix A contains the quantitative and qualitative questionnaires. These questionnaires consist of household level and village level questionnaires for each State.

An overview of data can be found in Appendix B. In Section 1. an overview is given of villages where Hill Resource Management Societies are active in Haryana and villages where *chakriya vikas prenali* is active in Jharkhand and Bihar. Section 2. presents comparative data at the State level of the three case studies. Section 3. presents comparative data at the village level of the villages which I have covered during my fieldwork.

Chapter 2

Institutions, Common Pool Resource Management, and Game Theory

1. Introduction

Arrangements for resource management are generally sought either through the market or through the State. Institutional alternatives, like the self-organised management of natural resources at the local level has emerged as a third option. This Chapter sheds light on the role of institutions within economic analysis in general. Institutions and organisation are defined and changes in instutional settings can be expected on long time scales. More specifically, various levels of property rights and regimes are presented and compared. Two theories of measuring the potential success in a common-pool regime are applied to the three case studies in this Book. Concrete examples of including institutions into economic analysis can be found in the game theoretic literature. Especially the issue of moving towards the Pareto dominated alternative in the prisoner's dilemma has received a lot of attention in the literature.

Many examples have emerged of successful people's managed common pool resources (Ostrom [150, 152], Sengupta [182], Singh [192], and Wade [212, 213]). Many of these examples have also shown how to avoid a tragedy in the situation where the population is growing exponentially and the absolute number of people who live in poverty is still increasing (Duraiappah [47], UNEP [204]). What can we learn from the experiences in managing common pool resources? Can it emerge as an institution on a larger scale? To address these type of questions, a broad overview of institutions helps to understand why, how, and what kind of institutions emerge at the community level to manage common pool resources.

In mainstream economics, the institutional environment is generally assumed to be fixed. In Section 2., I give an overview of the discipline of the 'new Institutional Economics' in an attempt to incorporate institutional change into economic analysis. The literature is surveyed for finding theoretical approaches that explain the emergence of proper institutional forms to manage common pool resources (Section 3.). In Section 4., I survey game theoretic approaches towards institutions in general and common pool resource management in particular. The purpose of this Chapter is to create a conceptual framework for understanding village level institutions, which emerge to manage common pool resources through collective action and people's participation. Finally, this Chapter ends with concluding remarks.

2. Institutional Economics

This Section gives a general idea for incorporating institutions, organisations and institutional change into economic analysis. This serves as a tool for using the discipline of economics in analysing institutions at the village level, in the rest of this Book.

2.1. The Need for an Institutional Economics

Let us consider the emergence of institutions in the history of economic thought. Among the pioneers in economic theory, it was Adam Smith who was already concerned with the relationship between institutions and wealth (Eggertsson [49, page 224]). Among the neoclassical economists, there were only a few authors like Leon Walras and Paul Samuelson (Field [51, page 690]), who were trying to link institutions with economic models. But many others neglected institutions or assumed institutions (mainly of money) to be fixed. This simplification and omission paved the way for an elegant and rigorous approach to economics. The formal approach of the neoclassical economics, which is based on a set of rather rigorous assumptions, has recently received much criticism.[1] The criticism that the approach has become too limited, has drawn the attention of economists towards incorporating institutions into the framework of economic analysis. Furubotn and Richter give the following motivation for an institutional approach.

> The tendency to introduce greater institutional detail into the economic models has come about gradually over time because of the recognition that standard neoclassical analysis is overly abstract and incapable of dealing effectively with many current problems to theorists and policy makers. [60, page 1]

In addition, two more streams have emerged in response to the limitations of the neoclassical approach. They are:

- Game theory, and
- Evolutionary economics.

Out of these, the evolutionary approach is seen as the most promising approach to explain institutional change, claims Mirowski [136]. But also game theory has a good scope for incorporating institutions in the analysis, because it deals with the rules of behaviour, trying to rationalise optimal responses of players to these rules. Section 4. discusses game theoretic approaches in relation to institutional analysis.

Yet, there is no agreement on how to include 'institutions' into an economic analysis. Rutherford [172], for instance, points out that there is a seeming dichotomy between two

[1] This argument is supported by many new institutional economists like Coase [34, 35], Commons [38, 39], Eggertsson [48, 49], North [142, 143, 144, 145], and Williamson [216, 217, 218] among others. North [144, page 19] explicitly states the neoclassical behavioural assumptions.

streams in economic analysis for analysing institutions: the "old" and the "new" institutional approach. The line of thought of the "old" institutional economics originated by the works of 'Thorstein Veblen, Wesley Mitchell, John R. Commons, and Clarence Ayres' [172, page 1]. Followers of this line of thought are generally in favour of an economic approach which is less formal, more holistic, more behaviouralist, more collectivist, and more interventionist. Dugger [46] points out six view points of the old institutional approach: [2]

1. Existing power relations in the economy are appreciated.
2. Existing institutions are looked upon with scepticism.
3. Human thought and action are seen as a dichotomy between other-help and self-help.
4. The evolutionary approach to social provisioning is followed.
5. The economy is conceived to be a part of an evolving whole—of a culture.
6. Institutions are put in a never ending process of self-adjustment.

The second stream, which came about more recently, is the "new" institutional economics and is rather mild in criticising the neoclassical approach.[3] Contributors to this stream are large in number, among them are Coase [34, 35], Commons [38, 39], Eggertsson [48, 49], Furubotn [59, 60], North [142, 143, 144, 145], and Williamson [216, 217, 218]. The new institutionalists do not reject the neoclassical economic approach, but they argue that the existing theory should be enriched by focusing more on institutions. The classical approach was not wrong, just incomplete. Followers of this line of thought are generally in favour of an economic approach which is more formal, more individualist, more rational choice, more evolutionary or invisible hand, and less interventionist.

In spite of different orientations in the old and the new institutional economics, they are not completely each other's opposite. The old institutionalists complain about the partiality of the new institutional approach. The other way around, new institutionalists try to incorporate more and more components of the old institutionalists into their approach. Rutherford [172] calls for an elimination of the mutual mistrust between both visions, to start a more constructive debate.

Alchian [3] is known as a pioneer of the new institutional economics. His analysis typically has very few assumptions. He shows that the idea of individual profit maximisation becomes a meaningless exercise in a situation of imperfect foresight and incomplete information.

Still, there are risks in relaxing these 'rigorous' assumptions of the neo-classical economics, in the move towards a new Institutional Economics. On the one hand, staying too close to the assumptions of the neoclassical economics may lead to *path dependency*, which led many scholars to 'a history of a cyclical rediscovery of the obvious' (Eggertsson [49, page 224]). On the other hand, by getting away from these rigorous assumptions there is the threat of being caught in the *open-field syndrome*, when research becomes too diversified. This can hinder the development of a body of theory, refined and extended by successive generations of scholars [49]. A real research challenge is to find a compromise between these two extremes.

[2]See Dugger [46, pages 424–428] for a further elaboration on the elements of what he calls institutionalism.

[3]Eggertsson made a distinction between the neo-institutional economics, based on optimising models and the new institutional economics, based on ideas of bounded rationality. This distinction is, however, difficult to maintain, according to North [144], due to the emergence of optimisation models with bounded rationality.

2.2. Institutions versus Organisations

The term 'institutions' has been used at various occasions in this Book, but what does it really mean? Many interpretations of institutions emerged. Authors defined institutions so that it could be fitted in their analysis. Ruttan, Hayami and Kikuchi [81, 173], for example, have a focus on institutions in a rural context, while Ostrom [149] elaborates more on formal rules. An attempt to give a broad definition is simply absent in many writings. In addition, social scientists tend to use the words institutions and organisation interchangeably. But institutions and organisations do not necessarily coincide. How exactly they differ was clarified by Uphoff [206]. He gave a very precise definition using sociological terminology.

> *Institutions*, whether organisations or not, are complexes of norms and behaviours that persist over time by serving collectively valued purposes, while *organisations*, whether institutions or not, are structures of recognised and accepted roles. [206, page 614]

Therefore, institutions that are seen as a specific complex of norms and behaviours, have a role to determine: what a person is supposed to do and how to do it.

For the present analysis, it will be more useful to define institutions and organisations in economic terms. North [144] devoted a book to explain the interaction between institutions and organisations. Based on his work, Eggertsson gave the following definitions:

> Institutions are defined as effective formal and informal rules that constrain individual behaviour, ... the institutional environment varies with a person's position in society. ... The term organisation refers to a set of actors who cooperate or act jointly in production and also to the rules that they play by. [49, pages 225–6]

The term *institutional environment* in this definition calls for certain general social rules. This should be distinguished from *institutional arrangements* which are particular organisational forms (Rutherford [172, footnote 1]). So, an organisation is defined in terms of people, while institutions are defined in terms of people's mind set. Consider the following quotation:

> We cannot see, feel, touch, or even measure institutions; they are constructs of the human mind. (North [144, page 107])

North distinguishes clearly between formal and informal rules. Formal rules are humanly devised, such as the constitution, laws, contracts and property rights. Informal rules are conventions and codes of behaviour like sanctions, taboos, customs, traditions, norms and values. These formal and informal rules make sense only when they are enforceable. To make institutions perform their task, a mechanism is needed to punish people who do not follow the prescribed behaviour. The strength of an institution is qualified by the effectiveness of the rule configuration; whether formal or informal rules can be enforced.

But what do the institutions do? Where are they needed for? Institutions can be seen as *constraints* which shape human interaction. Institutions are there to *reduce uncertainty* by providing a structure to everyday life. They make things easier. This is the *enabling* component of institutions. The presence of institutions gives a safe feeling.

A particular set of rules of the society induce the creation of organisations. Those organisations which work better under the rule configuration or institutional environment have a better chance to survive. A particular institutional environment induces incentives to create particular kinds of organisations. However, this does not necessarily mean that the institutional environment is efficient or that it will eventually become efficient. Even the use of the word efficient is problematic here. It may be efficient for some organisations, but it may be counterproductive for others. Incentives generated by an institutional environment can be favourable for some organisations, but unfavourable for others.

Just as some organisations can perform better than others, it is also useful to see institutions as matters of degree (Uphoff [206, page 614]). Institutions can be institutionalised to a certain extent. For example, the *van panchayats* (see Chapter 9) are fully institutionalised because people have formal rights in their communal forest and they are supported by the State since 1931. On the other hand, the rights of the members of Hill Resource Management Societies (see Chapter 7) are still not clear: they are awaiting an institutional status. Uphoff clarifies this distinction as follows.

> "Institutionalisation" is a process, and organisations can become more or less "institutional" over time to the extent that they enjoy special status and legitimacy for having satisfied people's need and for having met their normative expectations over time. [206, page 614]

The main objective of this Section was to elaborate on the role of institutions within the literature. Intuitively, institutions can repeatedly provide certain services to different people, irrespective of their initial intention. The replicability of this service is an essential characteristic which makes it an institution. For example, consider a specific case in Purulia district of West-Bengal. In Ulidih, a tribal village, the villagers were organised spontaneously by one of the villagers so that they became independent of outside money lenders (Lise [109, 110]). Here the altruistic behaviour of the village-leader constituted a service to the other villagers, which led to collective action through organising the villagers. This is an example of an organisation which is not an institution, because the service is not provided repeatedly. Once it was brought about, there was no need to do it again for this village. Of course, if this case would be taken as a model and the same would be replicated in some neighbouring villages, it can lead to a process of institutionalisation in replicating the service of voluntarily organising the villagers. Only then the process becomes an institution.

2.3. Institutional Change

Institutions, as discussed above, are generally not fixed. Over time changes are likely to occur. Institutions can either be created willingly or evolve autonomous over time. In addition, it is also possible in a feedback loop that organisations, which evolve from an institutional framework, can induce institutional change, for instance, when there are too many unfavourable rules. This can induce abrupt changes in formal rules. In an extreme situation, it can even lead to a revolution. Then, changes in formal rules are discontinuous. However, these abrupt changes shall not immediately change the unwritten informal rules. Informal rules tend to change incrementally over time. It is very difficult to dictate changes in informal rules. Ostrom expresses this as a conflict-invoking process.

> Instead of presuming that optimal institutional solutions can be designed easily and imposed at low cost by external authorities, I argue that "getting the institutions right"

> is a difficult, time-consuming, conflict-invoking process. It is a process that requires reliable information about time and place variables as well as a broad repertoire of culturally acceptable rules. [150, page 14]

For example, the caste system in India has been formally abolished since independence, but it is still playing an important role in the (rural) society. [4]

The direction of institutional change is generally perceived to be *path dependent* (North [144]), where the existing institutional environment is decisive for the ability to change rules. This does not mean that an institutional environment is unable to improve, but it is possible that some institutional environments provide the right incentives for growth, while others preserve economic stagnation.

The previous Section made a distinction between formal and informal rules, which helps explaining *institutional change*. While formal rules can be changed abruptly, informal rules change incrementally. While changes in formal rules can be created and intended, changes in informal rules tend to evolve unintentionally. Rule changes are generally more effective when the proposed new rules are enforceable or are supported by a large fraction of the society.

While changes in formal rules can induce changes in informal rules, there are many other factors that can lead to other changes. This makes it very difficult to control the change process of informal rules. Consider, for example, the case of forest management in Uttaranchal (Chapter 9). The change in policy of the British rulers to reserve all forests outside the village boundaries, by denying the people all rights in those forests, was easily formulated. It required only a change in formal rules. However, enforcement of these rules turned out to be problematic. It led to an unexpected large-scale resistance of the local people. The formal rule-change was not compatible with inducing the required change in informal rules. The strength of traditional norms and values on their turn induced a change in formal rules, when the colonial rulers had to return a part of the forest to the people. This case shows the interrelatedness between formal rules, informal rules and enforcement.

The functioning of institutions is generally well understood, but a formal framework to analyse institutional change, is far from being completed. Institutions do not only differ from case to case, but depend also on other institutions. More case studies have to be undertaken to understand under what circumstances which institutional changes can be expected. The lack of a formal model may seem discouraging, but institutional analysis is a promising contribution to the social sciences in general and the economic discipline in particular. It

[4]The caste system can be divided into five hierarchical layers:

1. The *Brahmin* or priest caste, also known as *Pandits*, is the highest caste.
2. The *Ksatriya* or fighters caste, also knows as *Rajputs* and *Jats*, is the second highest caste.
3. The *Baiyas* or skilled workers caste, for example *gujjars* and *Kahars*, is the third level in the caste system.
4. The *Sudra* or unskilled workers caste, also known as *Lohar*, *Bhanjda*, *Harijan*, and scheduled castes (SCs) form the lowest level in the caste system.

Finally, there are also people falling outside the caste division. These are known as scheduled tribes (STs). Many of them are found in Jharkhand, where each village has a different tribe, often including a different dialect or language.

5. The STs are in a way even lower than the lowest caste in the caste system.

A detailed numerical overview of the various castes encountered during fieldwork can be found in Table B.6 at page 189 in the Appendix.

provides an intuitive way of explaining the enormous variation in economic performance in different countries, regions, communities and so on.

3. Common Pool Resource (CPR) Management

This Section assesses theoretical approaches towards explaining the emergence of common property regimes. It deals with general approaches to managing a common pool resource (CPR) for analysing specific situations of forest management in India. First, I distinguish between property regimes and link these with degrees of 'ownership'. Second, I give a motivation for analysing common pool resource situations. Third, I conclude this Section with an overview of two theories to explain variation in success of common pool resource institutions.

3.1. Property Rights

A very important component of an institutional economic approach is to consider property rights.[5] The term *common property resource* has been treated interchangeably as open access and as common property regimes. In order to end this confusion, Bromley [26] made a distinction between *rights* and *property*. Moreover, he stressed the importance of a *benefit stream* in relation to an *object* to separately define property:

> Property is 'the *benefit stream that arises from that object or that resource*' [26, pages 92–93]

He continues to define rights more conventionally:

> *A right is the capacity to call upon the collective to stand behind one's claim to a benefit stream (that is, to one's property).*
> Rights can only exist when there is a social mechanism that gives duties and binds individuals to those duties. [26, page 94]

Property rights are institutions that regulate the control of all kinds of resources. Each member has the *right* to appropriate the resource, while the non-members have the *duty* to refrain from appropriation. Additional institutions are required to enforce rights like law and order to support a particular allocation of property rights.

When one talks of property rights, it is very useful to distinguish between four property regimes.[6] This distinction is necessary to avoid confusion between common property (*res communes*) and open access (*res nullius*). The four property regimes are:

a. A private property regime;
b. A common property regime;
c. A state property regime;
d. An open access or a non-property regime.

[5]Furubotn and Pejovich [58] give an extensive overview of the literature on the rise of property rights in economic analysis as a reaction to the neoclassical approach until 1972. De Alessi [41] extends this overview with more recent contributions until 1980.

[6]Ciriacy-Wantrup and Bishop [33] were the first to make this distinction. See also Bromley [25, 26, 27].

Table 2.1. A comparison between use-rights in different types of management regimes.

right-holder	property regime	rights	Example
individual	private	owner	a car
group	private	owner	a firm
group	common	{ proprietor claimant	a communal forest
public	state	authorised user	a road
"nobody"	open access	user	a grazing-pasture

To explain the differences between all the property regimes, it is also useful to follow the distinction of five degrees of 'belonging', as made by Schlager and Ostrom [179]. They are:

i. A **user** has no right at all, the resource is just used;
ii. An **authorised user** has the legal right to use the resource;
iii. A **claimant** additionally has also voting-rights in management;
iv. A **proprietor** additionally has also a share in deciding whom to exclude;
v. An **owner** can even sell or lease his/her collective-choice rights.

This distinction can prevent a wrong use of the word 'ownership.' For state property, ownership is either of a different dimension or it does not exist at all. Here, the highest level of ownership is an authorised user. A common property regime also has no particular owner: It is looked after by a particular group of people, but they do not enjoy the same privileges as a private owner. Here, the highest level of ownership is a proprietor. The legal protection of rights in a common property regime is generally weaker than for a private property regime. Unwritten rules like social sanctions are required for effective management of a common property regime. The link between rights and management regimes is displayed in Table 2.1. The rights are given in a descending order.

However, the link between property regimes and rights is generally not a one-to-one correspondence as it is shown in Table 2.1. For instance, it depends on the strength of a local institutions for managing resources through a common property regime, whether right-holders are authorised users, claimants, or proprietors.

The main distinction between group-private property and common property is that in the former case the common resource can be subdivided into individual owners; this division is not possible in the latter case. Consider for example the case of Jharkhand (Chapter 8), where the pool is considered as a group private property regime, because pooled private land can be withdrawn from the common pool. When this happens in Uttaranchal (Chapter 9), for instance, it is called encroachment.

For open access, the group of users is not properly defined. This class of management regimes is most liable to mismanagement and under threat of degradation. Due to the creation of a confusing situation, nobody wants to bear, or everybody denies, the responsibility. Historically, centuries ago, some common pool resources were open access regimes. The incentive to deplete the resource was low, because of low population pressures. A (costly) protection mechanism for enforcing well-defined property rights was not yet required. The situation changed when the population increased. A reduction in the quality of the common pool resources enhances the need for protection. Moreover, scarce resources should get the highest priority in protection. This protection is more difficult to enforce when the population increases even further.

The idea of a regime can also be used to distinguish between transient and permanent property rights. Dependent on the institutional environment, the management regime can be more or less *permanent*. However, management regimes tend to be *transient* in the long run, because the way of managing a particular resource can change over time. It may be sold from one individual to the other, but it can also be acquired by the State. To call a particular resource a *common property resource* means that it is managed at that particular moment by a well-defined group of people where the use-rules are defined as well. This management regime does not need to be stable. An institutional breakdown can make the same resource open to all. Since resource regimes need to be permanent, common pool resources cannot be defined in absolute terms. Bromley states:

> I suggest that there is no such thing as a common property *resource* — there are only common property regimes over certain natural resources in specific settings, and at particular times. [26, page 96]

The nature of property regimes is also revealed in an historical reconstruction of management regimes. The analysis of changes in property rights is not a typical economic exercise according to Ensminger and Rutten [50]. The objective is not to optimise but to suffice in a community:

> "as long as it works, do not fix it." Ensminger and Rutten [50]

They use an anthropological approach to analyse changes in property rights of an African pastoral commons in history. They observed two trends. First, a shift from private to common property, but as time went on, it shifted back from common to private property.

A way to include the changing nature of property regimes in an economic analysis, is by employing transaction costs. Field [52] distinguishes between *transaction costs*, the cost of making and enforcing agreements among users, and *exclusion costs*, the cost of excluding non-users from a resource. When the exclusion costs are too high it can pay to cooperate and choose for pooling private property in a common property regime, where the exclusion costs are much lower. But when the transaction costs become too high, privatisation is more beneficial. This shows that, historically, it was sometimes more profitable to privatise (when exclusion costs were low) and at other times it was more profitable to move towards common property (when transaction costs were low). This is an antagonistic process of change in property rights.

The emergence of common property regimes can also be explained by using the lab. Experimental laboratory studies, give further insight in the psychological behaviour of the human mind with respect to property regimes. The general argument is that in open access regimes rents dissipate completely, due to lack of responsibility. Ostrom [153] shows that this does not hold for common property regimes. Walker *et al* [215] show that learning due to repetition reduced the level of rent dissipation. Continuing to work with the same experimental framework, Gardner *et al* [62] explore the consequences of communication and sanctioning mechanisms among the subjects in a test. They arrive at the conclusion that sanctioning mechanisms alone even decrease the status quo, due to overuse of the mechanism. But communication improved the yield from common pool resources. The sanctioning mechanism should only serve as a credible threat to encourage a better yield. These experiments show the importance of communication to bring about participation. I get back to this issue in Chapter 5, which links the result of the estimated games to the outcomes of experiments in the lab.

The case-studies which are presented in Chapters 7–9 embody three different property regimes. In Haryana, the community forests are still under strict State control. In Uttaranchal successful *van panchayat*s have established community norms to enforce proper management. In Jharkhand the common pool created from private property has no ownership responsibility to the State. Their common pool, is legally protected: The State cannot put any claim on their pooled private lands. In a dynamic sense, there is a transformation taking place in Jharkhand from private to common property, in Haryana from state to common property, while the transformation from state to common property has already been completed in Uttaranchal, but it is reversing for communal forest where the *van panchayat*s are not functioning well.

3.2. Measuring Success in CPR Management

The argument in this Book makes a case for creating common property regimes, when other management regimes for common pool resources do not lead to preservation of the common pool resource. While the management regimes of all the three case studies are different, I expect that they will be managed as a common property regime in future. This Section focuses on the following questions:

- How can we judge the success of community managed common pool resources?
- Which factors affect common pool resource management positively or negatively?

Ostrom [150] shows through a number of case-studies that common pool resources are managed most successfully when all of the following eight design principles are present.

1. Clearly defined boundaries of the common pool resource.
2. Suitable use rules and rules for contributing private resource.
3. Members can influence the rules for managing common pool resources.

When these three components are present, the management rules are well-formulated.

4. The agreed upon rules are monitored properly.
5. Sanctions are proportional to the kind of violation.

When the last two components are added to the first three, rules are very likely to be enforced. The five factors, mentioned above, lead to well-formulated rules with an adequate enforcement mechanism. All these five factors relate to management of common pool resources at the field level. It also matters how the local setting is linked with outside organisations. She further adds the following questions.

Table 2.2. The presence of factors which enhance institutional success, according to Ostrom.

	State	Haryana	Jharkhand	Uttaranchal
1.	Clearly defined boundaries	yes	yes	yes
2.	Suitable rules for contributing and benefiting	yes	yes	yes
3.	Members can influence decision rules	yes	yes	yes
4.	Monitoring of rules is done properly	weak	yes	weak
5.	Proportional sanctions	weak	weak	weak
6.	Local court for resolving conflicts	weak	yes	weak
7.	Self-organisation permitted by the State	yes	yes	yes
8.	Part of larger system with various levels	no	no	yes
	Institutional performance	fragile	fragile	fragile

6. Are the rules interpreted equally and equitably? Is there a resolution mechanism, like a local court, for conflicts resulting from different interpretations and/or violations?
7. Is self-organisation permitted by the government?
8. Is the common property regime a part of a larger system? Is the larger system acknowledged by the government? [150, pages 88–102]

Institutions become more fragile whenever one or more components are missing from a particular practical situation. The extent to which most of these principles are met depends, among other factors, on the attitude and the reaction of the community on the common pool resource situation.

Table 2.2 shows that all the three studied situations have well-defined management rules, as the first three design principles are present. The score on the fourth design principle is mixed. I consider it weak in Haryana and Uttaranchal, where there is not always a guard and social fencing is also not always strictly applied. Moreover, meetings are not very regular, namely monthly in Haryana and quarterly in Uttaranchal. In Jharkhand the situation is different. There, regular fortnightly meetings are held, the pool is created on private land, and social fencing is strictly followed. Because of that reason a higher score has been assigned.

While sanctions can be applied to rule violators within the organisation, there are often violators from other villagers and communities. It is more difficult to punish them, and a legal system is needed here. Because of that reason the score 'weak' has been assigned to the fifth design principle. Sanctions can be assigned in various ways, like confiscating an axe in Uttaranchal or a monetary fine in Haryana. In Jharkhand rule violators are tried in

Table 2.3. The presence of factors which enhance institutional success, according to Wade.

	State	Haryana	Jharkhand	Uttaranchal
A.	Property regime	state	private (group)	common
B.	Investment in exclusion	low	high	medium
C.	Importance of the CPR	low	high	medium
D.	Group composition	heterogeneous	mixed	homogeneous
E.	Reduced role of State	no	yes	medium
	Equality index:	0.59	0.69	0.86

kind of a local court, during one of the frequent meetings. This explains the choice for weak in Haryana and Uttaranchal and yes for Jharkhand for the sixth design principle.

One aspect is common among all the three studied situation, namely that self-organisation is permitted by the State (design principle seven). While Haryana and Jharkhand are initiatives at a relatively small scale with around 45 organisations in place,[7] the situation in Uttaranchal is quite different. The origin of *van panchayats* is much earlier and started even before independence. In the relevant districts about 50% of all villages have a *van panchayat*. Indeed the more than 4500 *van panchayats* are part of a larger system with various levels, as shown in Figure 9.1 and Figure 9.2 in Chapter 9.

In principle, the three institutions as studied in this Book are effective. In reality this effectiveness is not always achieved. The effectiveness of Ostrom's eight design principles, as shown in Table 2.2, differ from village to village. Moreover, joint forest management in Haryana has not yet led to a common property regime of the communal forests, because the lease rights of the forests are still in the hands of the State. Pooling in Jharkhand has not yet led to a common property regime of pooled land, because people can still withdraw their privately held land from the pool. In Uttaranchal, the lease-right is in principle in the hands of the community, which makes it a common property regime.

Wade [213] confines his attention to the emergence of local organisations for managing a common pool resource. He is looking for factors which explain the differences in success between organisations for managing a common pool resource. He distinguishes between the following five factors.

A. The way *property rights* are attached to resources.
B. The amount of investment in technology for *exclusion* and detection of violators.
C. The *importance of the common pool resources* for the users.
D. The *composition of the group* of users with respect to size, power relations, amount of communication and reputation.
E. The extent towards which *role of the State* is reduced towards the users.

[7]For an overview of the Hill Resource Management Societies in Haryana see Tables B.1–B.2. For an overview of villages under *chakriya vikas prenali* in Jharkhand and Bihar see Tables B.3–B.4 in the Appendix.

Wade's classification is very useful for comparing the three case studies of this Book. The presence of Wade's factors are shown in Table 2.3. The property rights are essentially different in each case study. The amount of investment in protecting the forests also differ from state to state. In Jharkhand, the highest investment took place: the people who work in the nursery also protect the common pool resource. In Uttaranchal, each village has at least one formal guard. While, in Haryana, the investment in protection is rather low: in many cases there is not even a guard. The importance of the common pool resource differs also from state to state. It is considered on average most important for the people in Uttaranchal. The composition of the group is very case-specific and it is difficult to generalise this for the three institutional settings.[8] The role of the State also differs, because the property regimes are intrinsically different. Pooling in Jharkhand is most favourable to Wade's factors. This suggests that Jharkhand is the most participatory oriented institution. Also, the payoffs from participation are the highest in Jharkhand (see Table B.8 at page 191).

In Part III of this Book, I give an analysis of the impact of three different property regimes on the management of one specific common pool resource, namely forests. In many cases of empirical interest, common pool resources cannot be managed through the State and the market alone, and supporting institutions need to be present as well. However, State support is important for proper functioning of common pool resource institutions. The success of a common pool resource institution depends crucially on the amount of State support and on how well the market is functioning.

3.3. The Common-Pool Resource Situation

Forests are typically an example of a common-pool resource. This Section treats main insights from the literature on common-pool resources. The problem of managing a common pool resource, or simply the CPR problem, has become a topic of debate after the appearance of Hardin's [74] 'tragedy of the commons'. To analyse the common pool resource problem, it is useful to distinguish between the common pool resource situation and the common pool resource dilemma. In a CPR situation there are multiple individuals who are using a common pool resource. This becomes a CPR dilemma when the following conditions hold simultaneously:

1. Individual optimal strategies lead to suboptimal outcomes for the group, due to uncoordinated behaviour.

2. There are institutional feasible alternatives, where collective optimal outcomes can be achieved (Ostrom *et al.* [152]).

The common pool resource situation is generally not a dilemma, as Hardin presumes. The three case-studies in Chapters 7–9 are all examples of common pool resource situations only. common pool resource dilemmas existed in the past, but they were resolved through institutional change. The historical processes which led to the institutional change are briefly narrated below, but the focus is on the stability of the present institutional set-up in the three common pool resource situations.

[8] See Table B.7–B.8 in the Appendix for details on heterogeneity. The equality index, as calculated with formula A.7 on 169, also supports the division of the group composition in Table 2.3.

Initially, the solution to common pool resource dilemmas was sought in either full privatisation of property rights or full centralisation of the power to the State (i.e., government) (Hardin [74]; Hardin and Baden [75]). On the one hand, many common pool resources are not divisible, like sea-fisheries and the ozone layer. In such a situation privatisation is simply impossible. Furthermore, division may not be desirable for optimal use of the resource units because the newly allotted owners simply do not have the private resources to regenerate these already degraded resources. On the other hand, solutions were sought through the State by using taxes or subsidies. However, state property regimes are also degrading. For instance, in Uttaranchal the State was unable to manage the forests solely as a state property regime. People's protests led them to formulate the law on *van panchayats* where people are fully responsible for managing their 'communal forests'. The State, generally, fails to appreciate the ability of people to participate. However, as long as the people are not involved in regeneration processes initiated by the State, the people will not take any responsibility. Many authors looked for the possibility of other solutions (see for instance: Wade [212, 213]; Larson and Bromley [107]; Ostrom [150]; Stevenson [199]; Anant [6], among others).

A clear incentive for self-organisation among tail and head enders in an irrigation system was found by Ostrom and Gardner [151]. Statistical evidence on data from Nepal supports that self-organised irrigation systems work better than those which are organised by the government. However, success in self-organised groups came about after a struggle. Initially, self-organisation led to conflicts and confusion, but when the initiative to self-organisation was accepted by some villagers, other villages followed as well. This ultimately led to a management system for the whole irrigation channel.

Thompson and Wilson [201] argue for a proper mix between private and common property regimes. The latter are more adaptive to environmental variability. Nomads prefer to wander from place to place as the season changes. They take the opportunity to use grazing pastures in distant mountains. A common property regime can also reduce the risk of encountering a degraded resource, because the productivity can change from place to place, even within the resource.

To sum up, the appropriate management regime of a common pool resource is typically case-specific. The three case-studies presented in Chapters 7–9 provide clues for different solutions for three intrinsically different management regimes.[9]

It is also interesting to see the level of people's participation in forest management as social capital (the information, trust, and norms of reciprocity inherent in one's social network (Woolcock [219])). Following the theory of social capital (see for instance Coleman [37]; Fukuyama [57]; Castle [29]; Woolcock [219]), villagers themselves are not aware how their level of participation can influence the performance at the village level. The emergence of social capital at the village level is generally unintentional. Here, we are interested in identifying villages with a high endowment of social capital; where the level of people's participation is highest.

[9]Dietz *et al* [45] provide a general discussion on the interlinkage between various solutions to the challenge of governing the commons.

4. Game Theoretical Approaches

This Section gives an overview of game theoretical approaches in explaining when and why people cooperate. Many different approaches exist within game theory. Extensions of the game theoretical framework to incorporate norms can also be found in the literature. For instance, informal rules (norms) can be incorporated in complex game theoretic models, as long as rule changes are exogenous (DeMarzo [42]). I also discuss earlier game theoretic models of the common pool resource situation. This Section serves as a motivation for the noncooperative game model in Part II of this Book.

4.1. General Game Theoretical Approaches

The general objective of game theory is to find which combination of player's strategies leads to an equilibrium. In a simplified game where players can choose between cooperation or non-cooperation, it can be verified under which conditions an equilibrium of mutual cooperation can be attained and sustained. This also implies that a game theoretic framework can be a useful tool to study whether there are incentives for people to participate in forest management.

A central result in non-cooperative repeated game theory is the Folk theorem, which states that

> 'any feasible individual rational payoffs can arise in an equilibrium if the players are sufficiently patient' (Fudenberg and Maskin [56]).

More specifically, it is always possible to sustain a higher payoff than the minimax payoff when the discount factor is sufficiently large in an infinitely repeated game. The minimax payoff is the lowest payoff a player can ascertain, irrespective of strategies of all other players. Fudenberg and Maskin [55] show that a similar result can be derived for finitely repeated games with incomplete information.

An application of the Folk theorem can be found in Green and Porter [67, 163], who consider changes in payoffs in an n-person infinitely repeated stage game (allows for changing payoffs) due to non-observable exogenous shocks. Their model consists of an n-firm cartel. They show that collusion can lead to an outcome above the Cournot-Nash equilibrium minimax strategy. They derive this outcome by using finite period trigger strategies, similar to the ones formulated by Friedman [53]. Rotemberg and Saloner [168] use a similar framework, but their payoffs change through observable exogenous shocks.

Abreu *et al.* [1] extend the model of Green and Porter, such that the past actions of *other* players are not observed. They assume that a person only observes own rewards and own actions of the past. The data which I derived during my field survey in rural India in 1995–1996 are fitting in the Abreu-model to the extent that I assume that the farmer can only remember benefits which accrued to him/her over the last year. In Chapter 5, I analyse a simplification of the Abreu-model, with two persons and two strategies. This simplification makes it possible to estimate the game numerically (Chapter 4).

Game theory has been a useful device to relax several assumptions of the neoclassical economic theoretical framework, while maintaining mathematical rigour. Game theory has led to a variety of models. On the one hand, in cooperative games players can discuss their intentions. The focus is mainly on the stability of a coalition. On the other hand, in non-cooperative games, it is generally assumed that players are not able to discuss their intentions. Among others, four major groups can be identified in non-cooperative game theory.

1. Normal form games.
2. Extensive form games.
3. Repeated games.
4. Evolutionary games.

The simplest game is the normal form game, which is played only once: a one-shot game. This game can be represented by a payoff matrix when the number of players and strategies is finite. A more complex game is the extensive form game which can be represented by a game tree. The advantage of a game in extensive form is that a variety of rules can be explicitly included in the model.[10] The simplest example of a repeated game is obtained by repeated play of a (fixed) normal-form game. In evolutionary games the focus is on evolutionary stable strategies, rather than on players. Greenberg [68] started the challenging task to relax the assumption of no communication in non-cooperative games. His model allows for coalition formation in all of the above mentioned non-cooperative game representations. The requirement for his framework is that the so-called standard of behaviour is stable.

An application of the theory of social situations (TOSS) of Greenberg [68] to the issue of climate change can be found in Lise and Tol [126]. That paper shows that a game in TOSS is a useful tool in obtaining more insight into the underlying institutions of climatic negotiations.[11]

Recently, with the help of computers, it is also possible to compute games numerically.[12] Computation of games has helped in a great number of applications in industrial economics (Tirole [202]), where policy makers are interested in the possible consequences of competition.

4.2. Formal Approaches Towards Norms, Values and Customs

A direct link between institutional economics and game theory can be found in the following two papers where game theory is used as a tool to explain institutional change in the late medieval period: Greif *et al.* [69], Milgrom *et al.* [134]. However, the most troublesome factor of an institutional analysis is the persistence of informal rules such as norms, customs and conventions. These are also known as informal institutions. In this Section, I discuss

[10] See for instance Ostrom *et al.* [152]. They discuss the impact of rule changes with the help of a non-cooperative game model in extensive form.

[11] A general game theoretic analysis of leakages in greenhouse gas emission reduction can be found in Kemfert *et al* [95]. The impact of global warming on tourism is studied in Lise and Tol [124, 125], while the consequences of more severe droughts due to climate change is studied from a policy perspective in Ginasante *et al* [66]. A game on water management in the UK can be found in Lise and Bakker [120], while a game between farmers competing for damwater in Spain can be found in Lise *et al* [121]. Gerlagh and Lise [64] show, with the help of a numerically calibrated growth model, that an explicit inclusion of technological change can lower the cost of greenhouse gas emission reduction.

[12] See, for instance, Admundsen and Bergman [5], Lise *et al* [122], Lise and Linderhof [123], Lise [118] and Pineau and Murto [159] among others.

attempts to model informal rules. The main reference is to game theoretic models with exogeneously changing norms.

There is no consensus on whether informal rules can be modeled meaningfully with the help of game theoretic models. Field [51], for instance, challenged this possibility. He raises doubts about the possibility to arrive at a framework which can explain altruistic behaviour of individuals, and the evolution of norms and rules.

Schotter [180] points out where the neoclassical economic theoretical framework is insufficient in its description and how a description can become more thorough by also incorporating institutional aspects. He generates a framework for describing the evolution of norms in the society. He analyses an n-person prisoner's dilemma supergame (which is derived from the infinitely repeated prisoner's dilemma by considering expected behaviour) and he incorporates norms. These norms are modeled as a chance distribution on the expected behaviour of the other players. A player's decision is based on his common belief towards other players. The norm is adjusted in every consecutive period. He describes this adjustment process by a dynamic diffusion process which consists of a set of partial differential equations. Peyton Young [220] combines evolutionary game theory and non-cooperative game theory building further on the ideas of Schotter [180].

Nishijima [141] uses the bargaining non-cooperative game theoretical framework of Rubinstein to model norms. This is in the sense of Schotter [180], where norms are seen as a chance distribution on the strategies of the players. He characterises a unique equilibrium.

Individuals make decisions, having limited sources of private information, which are generally not common knowledge. Kandori [92] uses random matching games where players have a private information set which changes as a learning process, due to interaction with different persons. He incorporates changes in private information in his framework and models changing norms where one can observe the 'label' of the opponent. The norm suggests the optimal response to each label. This shows that it is possible to model changing norms within a game theoretic framework.

Okuno-Fujiwara and Postlewaite [148] provide a *framework* for modeling social norms into random matching games. The norm consists of two components: a status level and a social standard of behaviour. In the case of finite players, the proposed equilibrium can collapse when too many people decide to defect on the same moment: a revolution. A new norm shall emerge. It is to be seen whether this new norm is again stable or not.

Sethi and Somanathan [184] use evolutionary game theory to explain the impact of norms in the community and they endogenise the resource stock. They consider three types of players:

1. Cooperators;
2. Defectors, and
3. Enforcers.

Enforcers exist when the (neoclassical) assumption of self-interest is relaxed. The strongest equilibrium, which is asymptotically stable, consists of a blend of cooperators and enforcers. Then there are no defectors.

DeMarzo [42] considers the most abstract game of all. His model is an n-person, m-strategy repeated multi-stage game (allows for changing payoffs), where coalition formation and communication can take place. His model follows the theory of social situations as formulated by Greenberg [68]. The norm is a strategy which, once suggested by a leader and accepted as a 'focal point', the players should follow. This typically characterises an exogenous norm. He believes that this framework can be extended to model norms endogenously.

4.3. Possibility of Cooperation in a Repeated Prisoner's Dilemma

A common pool resource dilemma can typically be represented by a prisoner's dilemma, because the prisoner's dilemma shows a conflict between individual and collective rationality, which is the first requirement for a common pool resource situation to become a common pool resource dilemma. In addition, there are institutional alternatives to switch from the Nash equilibrium to the Pareto dominating outcome; satisfying the second requirement (see also Section 3.3.). In the prisoner's dilemma, the collective rational outcome is self-destabilising, because deviation is a dominating strategy. Even if the prisoner's dilemma is repeated finitely and the number of repetitions are common knowledge, it is well known from the literature (see, for example, Luce and Raiffa [128]) that the subgame perfect equilibrium of the game is mutual defection in each period, using the backward induction principle.

The discrepancy between the subgame perfect equilibrium of the finitely repeated prisoner's dilemma and the collective optimal outcome has been given much attention by several authors. Van Damme [207, page 170] refers to at least three formal models which eliminate this discrepancy. The possibility to accomplish cooperation in the repeated prisoner's dilemma can be divided broadly into three classes, following Seabright [181, page 119].

More complicated behavioural strategies than the standard trigger strategies can yield cooperation in equilibrium of the finitely repeated prisoner's dilemma.

Hirschleifer and Rasmusen [83] analyse an n-person prisoner's dilemma with ostracism. The punishment of a one period exclusion from the community can be severe enough to secure cooperation as an equilibrium outcome.

Matsushima [133] finds an equilibrium in random matching games, with the structure of a finitely repeated game, when players continue with cooperating partners and opt for a new partner through a random matching procedure as soon as the present partner defects.

Frohlich [54] uses a notion of fairness "you cut and I choose" to generate cooperation as an equilibrium outcome. The players make a choice first. Then their roles are randomised, so that they are not sure what they will get.

Hardin [76] shows that cooperation can occur by using the concept of majority rule, which he calls a Condorcet choice. He shows that a weak or strong majority who prefer cooperation as the outcome can always be formed. Hence, some cooperation can be realised.

An unknown termination time in the finitely repeated prisoner's dilemma makes mutual cooperation a possible equilibrium outcome.

Basu [13] argues that in practical situations, termination of the game is common knowledge, but the number of repetitions is unknown. For instance, if a game ends when one player dies, it is unrealistic to assume that the number of repetitions is common knowledge. In this situation, mutual cooperation is a possible equilibrium outcome.

Axelrod [9] invited scientists from different disciplines to design a strategy for playing a finitely repeated prisoner's dilemma where the termination time is unknown. A tit-for-tat strategy did very well: start with cooperation and do what the other player did in the previous period. It also won the tournament over all submitted strategies.

Samuelson [174] derives a cooperative solution for the prisoner's dilemma where the termination time is not common knowledge, but private information. Mutual cooperation becomes a possible outcome.

Incomplete information was used by Kreps *et al.* [100, page 246] to derive that the number of defections is bounded from above, with a little bit of (the right kind of) private information (Gibbons [65, page 142]). Bounded rationality, where a player does not necessarily choose the expected optimal outcome, can also lead to cooperation: by 'mistake' both players can start cooperating, but pure rational players never deviate from their dominating strategy.

Fudenberg and Maskin [55] analyse a situation where a player is sane with probability $(1-\varepsilon)$. A player is 'crazy' with probability ε and cooperates until the opponent defects, after which he also defects.

Other games can also rationalise mutual cooperation. Okada [147], for instance, shows that cooperation is possible in an n-person prisoner's dilemma in four stages.

1. Members decide if they want to cooperate;
2. Members bargain for an enforcement agency;
3. Members bargain for a proper punishment mechanism to eliminate the temptation to deviate, to be applied to all group members;
4. Members either come to an agreement or not.

In this game k members ($k \leq n$) cooperate and $n-k$ members do not. The four-stage game of Okada has not yet been analysed in a repeated framework.

Another way to rationalise cooperation in the repeated prisoner's dilemma is by considering *infinite repetitions*. This route is followed in Chapter 5. Again another possibility is to consider finite period punishments or by making the level of payoffs dependent upon the quality of resources. Chapter 6 explores this path.

5. Concluding Remarks

This Chapter has provided a broad overview of the literature on institutions, common pool resource management and game theory in particular. Especially the body of game theoretic literature is a promising direction for understanding common pool resource situations. This provides an important motivation for the study of one particular institutional setting: people's participation in forest management in India.

This Chapter started by distinguishing between institutions and organisations. A distinction was made between formal and informal institutions and it was pointed out that the threat of sanctions are required to make institutions effective. Social rules like norms are informal institutions and legal rules like the law are formal institutions. While formal institutions are generally malleable, norms are path-dependent, change incrementally, and the direction of change is generally difficult to control.

A distinction was made between different property regimes in connection with different degrees of ownership. The argument here was that there is no superior property regime, but in some cases common pool resources are better managed as a common property regime.

Furthermore, I discussed how game theory can accommodate institutions into their models. An important objective of game theory is to study the conditions under which mutual cooperation among players is likely to emerge. Therefore, game theory is seen as a useful tool for analysing people's participation in forest management.

In the literature a number of authors can be found that have designed game theoretic models including norms. These models can be useful to avoid the classical prisoner's dilemma which shows a conflict between individual and group behaviour. In addition to including norms into the model, I have discussed three other ways to solve this dilemma.

Part I of the Book is a prelude to Part II, as one of the purposes of the field survey was to find factors which enhance participation. This is verified in Chapter 3 by employing an econometric model and by undertaking a factor analysis on indicators of participation, which have been collected during my field survey. In addition, the simple two-person game which is central in Chapters 4–6 is also a useful tool for gaining additional insight into the likelihood of voluntary people's participation to emerge.

Part II

ECONOMETRIC AND GAME THEORETIC MODELS

Chapter 3

Factors Influencing People's Participation in Forest Management

1. Introduction

Forests used to be an important source of revenue for the government of India, which is no longer the case because of large-scale deforestation. Proper forest management is needed to regenerate degraded forests, yet the government is powerless when people refuse to participate. However, there might be conditions that are more conducive for people's participation in forest management and this Chapter draws lessons from practical settings in which people do participate.

Hardin's [74] tragedy of the commons suggests either state intervention or privatisation of property rights to preserve common-pool resources. Institution building at the community level for managing common-pool resources has emerged as a third possibility (Wade [212, 213]; Ostrom [150]; Bromley [25, 26]; Sengupta [182]; Ostrom *et al* [152]; Singh [192]). This third possibility has been applied in rural India to the case of forest management. A number of studies signaled the importance of people's involvement in forest management (Chopra *et al* [31]; Palit [154]; Poffenberger and McGean [161]; Sarin [177]). These studies show that in many institutional settings of rural India, forests are better managed when voluntary people's participation is secured. Hence, there exist many situations where people's participation is desired, and it is of considerable interest to find conditions under which voluntary participation takes place, which is the purpose of this Chapter. Moreover, this Chapter studies the qualitative and quantitative link between participation and socio-economic variables to explore under which conditions people are most likely to participate voluntarily in forest management.

This Chapter is organised as follows. Data that is gathered through interviews is analysed with two methods. In Section 2., a factor analysis on indicators of participation is used to identify different choice situations. General patterns that enhance participation are derived by multiple regression (Section 3.). This Chapter concludes by suggesting factors that could be part of a general strategy for successful forest management.

Table 3.1. Principal components of participatory indicators in Haryana and Jharkhand.

State	Haryana		Jharkhand		
Level of participation	Social	Economic	Social internal	Economic	Social external
Factor	1	2	1	2	3
Planting in the forest	0.383	0.044			
Contribution to forest/pool	0.226	**0.698**	0.172	**0.650**	–0.152
Benefiting from forest/pool	0.094	**0.832**	0.098	**0.808**	0.072
Ability to use the pool			0.082	**0.742**	0.299
Benefits from using the pool			0.156	**0.706**	0.233
Importance of meetings			**0.535**	0.065	0.338
Agreement with decisions	0.049	**0.876**	**0.832**	0.164	0.165
Attendance of meetings	**0.797**	0.195	**0.787**	0.342	0.033
Ability to influence decisions	**0.682**	0.416	**0.868**	0.122	0.087
Frequency of meetings	**0.792**	–0.126	–0.020	0.021	**0.568**
Interest in the meetings	**0.640**	**0.531**	0.292	0.141	**0.721**
Gain from meetings	**0.611**	**0.529**	0.271	0.251	**0.775**
Suggesting in meetings	**0.584**	0.395	0.280	0.415	0.172
Percentage of variance explained	45.1%	14.5%	36.0%	12.3%	10.0%
Number of observations	127		123		

Note: Numbers in bold face denote a dominating indicator (factor loading ≥ 0.5 or ≤ -0.5). An empty cell means that the observations on that indicator are missing.
Source: See text.

2. Factor Analysis: Dimensions of Participation

For measuring attitudes toward the village council, the respondents were asked a set of questions, which are interpreted as *indicators of participation*. These indicators are scaled as an integer value in a range from one to five, where one means total disagreement and five means total agreement with one particular aspect of participation with respect to the village council. This Section reports on the results of a factor analysis (Harman [78]) on these indicators of participation, to qualify and quantify the level of participation.

A factor analysis, which is a method for translating a large set of variables into a few independent choice variables, separates participatory indicators into a set of principal components, known as factors. Each factor represents an independent choice. As a rule of thumb, variables with a coefficient in absolute value above 0.5 are said to be dominating in a factor. Another rule of thumb is that all factors with an eigen value larger than one should be used in the analysis.

Tables 3.1 and 3.2 show the results, which are quite diverse for all the three States. The factor analysis yield two, four, and three factors in, respectively, Haryana, Jharkhand and Uttaranchal. Where it is difficult to understand why these numbers of factors are derived, we shall see that it has a clear interpretation: the number of factors tells us the dimensionality of participation in each State.

For the case of Haryana, we can see from Table 3.1 that the dominating variables in the first factor, which explains 45% of the variation, are all related to people's attitude towards the meetings. This is typically a social aspect of participation. The dominating

Table 3.2. Principal components of participatory indicators in Uttaranchal and all India.

State	Uttaranchal				All three cases	
Level of participation	Social internal	Economic	Social external	Benefiting	Social	Economic
Factor	1	2	3	4	1	2
Planting in the forest	0.048	**0.867**	0.000	0.035		
Contribution to the forest/pool	0.177	**0.617**	-0.024	*0.493*	0.184	**0.797**
Benefiting from the forest/pool	0.233	0.109	–0.074	**0.782**	0.111	**0.815**
Ability to use the pool						
Benefits from using the pool						
Importance of meetings	0.095	0.053	**0.885**	–0.115		
Agreement with decisions	**0.771**	0.239	–0.165	0.156	**0.580**	0.427
Attendance of meetings	**0.594**	0.421	–0.139	–0.026	**0.714**	0.229
Ability to influence decisions	**0.810**	0.180	–0.101	0.116	**0.798**	0.215
Frequency of meetings	0.275	0.143	*–0.474*	–0.415	**0.594**	–0.068
Interest in the meetings	**0.842**	–0.037	0.086	0.003	**0.804**	0.188
Gain from meetings	**0.815**	0.006	0.175	0.131	**0.775**	0.250
Suggesting in meetings	*0.488*	0.451	0.305	–0.274	**0.512**	0.309
Percentage of variance explained	35.4%	11.9%	10.7%	9.2%	45.6%	11.9%
Number of observations	135				385	

Note: Numbers in bold face denote a dominating indicator (factor loading $\geq$0.5 or ≤ -0.5). Numbers in italic face are almost dominating (factor loadings close to 0.5 and –0.5.) An empty cell means that the observations on that indicator are missing.

Source: See text.

variables in the second factor, which explains 14% of the variation, express contribution to and benefiting from participation as well as agreement to decisions. The interest to attend the meetings and the purpose it serves to the participants has again a high factor loading. While economic considerations are important in the second factor, two participatory indicators related to meetings are also dominant. These two participatory indicators relate to the acceptance of the meetings, whether they can conform themselves to the discussions in the meetings. The second factor represents people's economic benefit and contribution and their acceptance of the village council. This shows that participation in forest management in Haryana consists of two dimensions.

In Jharkhand, three factors are found and to each factor a very distinct meaning can be given. The first factor shows the extent people have appreciation for the meetings and respect for the decisions; whether they accept the local organisation or not. These are typical social considerations about how people appreciate the meetings, so let us call this factor internal social participation. It reflects to what extent people identify themselves with the village council; whether the council is internally coherent. The second factor shows the extent of contribution to and benefits from pooling, whether they are economically involved in the local organisation or not. It appears natural to label this factor as economic participation: the extent to what the village council enhances people's welfare. It reflects the impact of the village council on the villagers and their willingness to improve the forest through this village council. The third factor shows to which extent meetings are seen as frequent and useful; whether the village council is perceived as effective or not. This consists again solely of social considerations about acceptance of the meetings as a means to communicate; call this external social participation. This shows that participation in pooling consists of three dimensions.

In the case of Uttaranchal, the dominating variables in the first factor, which explains 35% of the variation, are all related to people's participation in evaluation and decision making which typically symbolises a social choice. It also symbolises the acceptance of the local organisation. The dominating variables in the second factor, which explains 12% of the variation, express people's contribution to the *van panchayat*, which typically symbolises an economic choice. Both the third and the fourth factor are dominated by a single variable. The third factor is dominated by the importance of the meetings and almost negatively dominated (factor loading –0.47) by the frequency of the meetings. This means that people who consider the meetings to be important also believe that the meetings are not held frequently; it is the reflection of people who are quite pessimistic about the present practice of the village council, which can only be improved by meeting more frequently. Note that six out of 11 indicators of participation are negative, but not dominating, supporting the statement of a negative attitude which is reflected in this factor. The third factor represents people's attitude towards the functioning of the meetings. The fourth factor is dominated by the amount of benefits from participation, while the amount of contribution (factor loading: 0.49) and the frequency of the meetings (–0.41) are quite important, but not dominating. This means that the persons who gain from participation and contribute more to the village council perceive the meetings as infrequent; they also are quite negative about their possible influence on the village council as indicated by four negative participatory indicators. The fourth factor resembles the level of benefits from participation. The factor analysis shows that participation in Uttaranchal is four-dimensional.

The first and the third factor of Jharkhand are joined in Haryana and Uttaranchal, with one exception in each case. In Haryana, acceptance of decisions has become a part of the second factor, while in Uttaranchal, the frequency of the meetings dominates the third factor alone. Contribution and benefiting of Jharkhand's second factor are also found in Haryana, but in Uttaranchal, contribution and benefiting are split over the second and the fourth factor.

Pooling of all observations leads to an average choice irrespective of the institutional set-up. This situation most resembles the case of Haryana. Two factors are found. In the first factor, all coefficients that are related to meetings dominate. In the second factor, both coefficients that are related to economic aspects of participation dominate. Hence, on the combined level we see a clear division of the participatory choice into two components where social considerations are most important; economic considerations constitute the second main important consideration. It is noteworthy that the first two factors at the State level and the combined level have a great resemblance: they represent similar choice situations in each institutional setting: (1) social participation and (2) economic participation.

This result is no real surprise because most indicators are on the social aspects of participation. The result could only be decisive if an equal number of indicators were considered.

Let the sum of the participatory indicators be called overall participation. In the analysis that follows all derived factors and the total sum shall be used as different representations of the level of participation.

3. Regression Analysis: Better Forest Management

This Section identifies under which conditions a person is most likely to participate in forest management. Links between several socio-economic variables and participation are found with the help of multiple regression analyses. This Section discusses these links. Table 1.1 (page 6) showed the general patterns for the three institutional settings. The following equation is estimated for 15 situations:

$$\begin{aligned} \theta &= \alpha + \beta_1 \text{RES} + \beta_2 \text{FORDEP} + \beta_3 \text{AVAGE} + \beta_4 \text{EDUS} \\ &+ \beta_5 \text{FMRAT} + \beta_6 \text{CONPC} + \beta_7 \text{INCPC} + \beta_8 \text{CAPPC} \\ &+ \beta_9 \text{CASGR} + \beta_{10} \text{RELNR} + \text{error} \end{aligned} \tag{3.1}$$

where θ is the level of participation, α is a constant, while the β_i's are the coefficients of socio-economic variables. Table 3.3 shows the meaning of these variables. Tables 3.4 and 3.5 show the results. Besides running regression model 3.1, the descriptive variables are also checked for multi-collinearity by excluding correlated variables. For instance, education of the interviewee is strongly correlated to the average education in the family. The latter is therefore excluded in all cases. It also led to state specific inclusions of variables, such as in the most cases the per capita income being correlated to consumption per capita led to a higher significance. The opposite is true for the case of Jharkhand and, hence, in that case the consumption per capita variable is used. Furthermore, it was found that capital per capita was strongly correlated to income per capita, consumption per capita and average age for the case of Uttaranchal, and is therefore excluded. Finally, the age and gender of the respondent are also excluded from the regression because they never became significant.

Note that the adjusted R^2 is low (≤ 0.19) and even negative in those cases where all considered descriptive variables are insignificant, i.e. internal social participation in Jharkhand and economic participation in Uttaranchal. Remark that a low adjusted R^2 is inherent to cross-section data, it is not caused by the sample size, and it suffices to interpret linkages with significant t-statistics. The variables that are not significant in the regression equations can also be interpreted; namely, that they do not influence the behaviour of interest as it is described by the dependent variable.

For a general interpretation of the regression outcomes of Tables 3.4 and 3.5, note that the factor on external social participation and benefiting in Uttaranchal is best interpreted as the higher the level the worse, because many participatory indicators in these factors are negative. The regression outcomes are quite diverse for the four institutional settings, but some general patterns are apparent. First of all, the level of resources is always positively linked to participation and significantly in nine out of 15 cases. This shows that participation is enhanced when the people perceive their resource as being of a good quality.[1] A similar conclusion can be drawn for forest dependence. This link is also positive in all cases and significantly so in six out of eight cases, meaning that high forest dependence stimulates people's participation in forest management. Better resources and increased dependency on the forest lead to a higher level of participation. This suggests that improving levels of resources strengthens people's participation. A higher level of forest dependence means

[1] In addition to a good quality of the forest it is more generally also important to have a good quality of resources, like the quality of water. The cost effective monitoring of water quality has been studied extensively for the Rhine basin by Behrendt *et al* [18], Lise [117] and Lise and Van der Veeren [127]

Table 3.3. Meaning of the included variables in the regression.

Variable	Meaning and definition
θ	The level of participation, based on the principal components or total sum of participatory indicators.[a]
RES	The level of resources, based on the principal component of three indicators: present quality, change in quality over a period of 10 years, availability of resources.[a]
FORDEP	Forest dependence: total use of forest goods, like fuelwood, fodder, timber, divided by total need per family.[b]
AVAGE	Average age in the family.
EDU15	Average level of education in the family of members older than 15 years.
EDUS	Years of schooling of the respondent.
FMRAT	Female-male ratio (number of female family members divided by the total number of family members times 2000).
CONPC	Consumption per capita.
INCPC	Income per capita.
CAPPC	Capital per capita.
CASGR	Caste group (higher number means a lower caste).
RELNR	Religion group (1=Hindu, 2=Muslim, 3=Christian).

[a] The level of participation and the level of resources are normalised between zero and one. The minimum value resembles the case where the respondents answered 'not at all' all the time; this value is not necessarily attained.

[b] Forest dependence is by definition normalised between zero and one.

Table 3.4. Links between socio-economic variables and levels of participation in Haryana and Jharkhand.

	Haryana			Jharkhand			
	overall	social	economic	overall	internal social	economic	external social
(Constant)	0.211	0.347⋆	0.147	0.517⋆⋆⋆	0.898⋆⋆⋆	0.523⋆⋆⋆	0.413⋆⋆⋆
	(0.155)	(0.146)	(0.168)	(0.119)	(0.086)	(0.141)	(0.095)
RES	0.247⋆⋆	0.0792	0.310⋆⋆	0.326⋆⋆⋆	0.0419	0.328⋆⋆	0.190⋆
	(0.093)	(0.0873)	(0.100)	(0.096)	(0.0697)	(0.114)	(0.077)
FORDEP	0.328⋆⋆⋆	0.207	0.269⋆⋆				
	(0.090)	(0.085)	(0.098)				
AVAGE	−0.00132	0.00114	−0.00306	0.00177	0.00068	−0.00020	0.00249
	(0.00221)	(0.00208)	(0.00240)	(0.00169)	(0.00122)	(0.00200)	(0.00135)
EDUS	0.00067	0.00695	−0.00622	0.00775⋆	0.00137	0.00343	0.00669⋆
	(0.00424)	(0.00400)	(0.00461)	(0.00338)	(0.00244)	(0.00402)	(0.00271)
FMRAT	−4.15E−05	−2.98E−05	−9.88E−06	6.27E−06	−1.22E−05	−2.14E−05	5.99E−05
	(6.59E−05)	(6.21E−05)	(7.15E−05)	(4.62E−05)	(3.34E−05)	(5.49E−05)	(3.71E−05)
CONPC				−6.72E−05⋆⋆⋆	−2.40E−05	−5.40E−05⋆	−1.68E−05
				(1.89E−05)	(1.36E−05)	(2.24E−05)	(1.51E−05)
INCPC	3.39E−06	−1.03E−05	1.82E−05⋆⋆				
	(5.72E−06)	(5.40E−06)	(6.21E−06)				
CAPPC	5.40E−08	1.58E−07⋆	−9.86E−08	1.72E−07	−7.56E−07	1.11E−06	−1.73E−08
	(6.70E−08)	(6.31E−08)	(7.27E−08)	(6.56E−07)	(4.74E−07)	(7.79E−07)	(5.26E−07)
CASGR	0.0299	0.0423⋆	−0.0110	−0.0083	−0.00816	−0.0075	0.00228
	(0.0176)	(0.0166)	(0.0192)	(0.0110)	(0.00796)	(0.0131)	(0.00884)
RELNR	−0.0015	−0.0643⋆	0.0890	0.0287	0.0116	0.0167	0.0124
	(0.0293)	(0.0276)	(0.0318)	(0.0201)	(0.0145)	(0.0239)	(0.0161)

Note: The value in the parenthesis is the standard deviation. $\star, P < 0.05$; $\star\star, P < 0.01$; $\star\star\star, P < 0.001$.

Source: see text.

Table 3.5. Links between socio-economic variables and levels of participation in Uttaranchal and all India.

	Uttaranchal					India		
	overall	internal social	economic	external social	benefiting	overall	social	economic
(Constant)	0.419⋆⋆⋆	0.466⋆⋆⋆	0.512⋆⋆⋆	0.514⋆⋆⋆	0.452⋆⋆⋆	0.388⋆⋆⋆	0.407⋆⋆⋆	0.494⋆⋆⋆
	(0.105)	(0.102)	(0.100)	(0.103)	(0.101)	(0.053)	(0.049)	(0.066)
RES	0.0639	0.0368	0.0048	–0.115	0.198⋆⋆	0.248⋆⋆⋆	0.139⋆⋆	0.250⋆⋆⋆
	(0.0713)	(0.0696)	(0.0682)	(0.070)	(0.0688)	(0.051)	(0.047)	(0.064)
FORDEP	0.194⋆⋆	0.225⋆⋆⋆	0.032	–0.120⋆	–0.0145			
	(0.061)	(0.059)	(0.058)	(0.060)	(0.059)			
AVAGE	–3.06⋆	–3.77⋆⋆	–0.45	3.79⋆⋆	–0.070	–0.23	0.06	–2.08
(10^{-3})	(1.40)	(1.37)	(1.34)	(1.38)	(1.35)	(1.04)	(0.96)	(1.31)
EDUS	0.00349	0.00496	–0.00264	0.00398	0.00174	0.00354	0.00503⋆⋆	–0.00352
	(0.00305)	(0.00298)	(0.00292)	(0.00300)	(0.00295)	(0.00206)	(0.00190)	(0.00259)
FMRAT	2.60	2.95⋆	–0.40	0.92	–0.82	0.64⋆	0.74⋆⋆	–0.06
(10^{-4})	(1.48)	(1.44)	(1.41)	(1.45)	(1.43)	(0.26)	(0.24)	(0.33)
INCPC	–0.12	–0.91	2.26	–0.26	–1.50	–5.05⋆	–6.94⋆⋆⋆	2.88
(10^{-6})	(2.55)	(2.49)	(2.44)	(2.51)	(2.46)	(2.33)	(2.16)	(2.93)
CAPPC						–1.41E–08	3.94E–08	–5.24E–08
						(5.44E–08)	(5.03E–08)	(6.85E–08)
CASGR	–0.0097	–0.0018	–0.0106	–0.0167	0.0136	0.0224⋆⋆	0.0287⋆⋆⋆	–0.0175
	(0.0214)	(0.0209)	(0.0205)	(0.0211)	(0.0207)	(0.0078)	(0.0072)	(0.0099)

Note: The value in the parenthesis is the standard deviation. $\star, P < 0.05$; $\star\star, P < 0.01$; $\star\star\star, P < 0.001$.
Source: see text.

that the people have a higher stake in the forest, which is reflected in their higher level of participation.

The indicator of the average age in the family is only (negative) significant in Uttaranchal in three out of five cases. This negative significance implies that the younger people in Uttaranchal participate most.

When we look at the indicator for education of the interviewee, three positive significant linkages are found, namely, for overall and external participation in Jharkhand and social participation in all three cases. This shows that when education is significant, it stimulates participation.

The linkage with gender is very interesting and shows a different pattern for each State. In Haryana the linkage is negative for all three kinds of participation, but never significant. In Jharkhand the linkage is positive for overall and external social participation, while it is negative for internal social and economic participation, but never significant. In Uttaranchal a high number of women in the family is linked significantly positive to internal social participation. The positive link between gender and participation in Uttaranchal indicates that more involvement of women strengthens the institution of people's participation in Uttaranchal. At the same time we find there an insignificant negative link with economic participation. Finally, when all three cases are joined, we find two positive significant links between representation of women in the family and participation, namely, for overall and social participation, while the link is insignificantly negative for economic participation.

Let us focus for the time being on women's participation and note that as compared to Haryana and Jharkhand, the productivity of cultivated agricultural land is highest in Uttaranchal. This is very remarkable because in Uttaranchal all agricultural land is rain-fed and agriculture is not mechanised, while in Haryana a part of the land is under irrigation and agriculture is partially mechanised. In the hills the main agricultural burden is taken by women, while this is generally not so in Haryana and Jharkhand. The higher productivity may be due to the smaller land-holdings in the hills, which is subjected to a high labour input, and in contrast to Haryana and Jharkhand, droughts are rare in the hills. This link suggests that women's involvement in agriculture has a strong positive influence on agricultural output. It is remarkable that we find a positive link twice between gender and participation for the combined sample.

The link between wealth (as measured by consumption, income or capital per capita) and participation is significantly positive in Haryana and significantly negative in Jharkhand. This shows that the participatory process in Haryana has the cooperation of rich people. This is not very surprising, since landowners have an interest in obtaining a right to damwater. In Jharkhand there is no such incentive and landless people cooperate because of the availability of labour in the pool of land. Furthermore, there is no significant link with wealth in Uttaranchal; there the people have other reasons to involve themselves in the process of participation.

The link between caste and social participation is significant and positive in Haryana. In Jharkhand and Uttaranchal it tends to be negative, but not significantly so. This shows for Haryana that people from a lower caste participate more. The same link is also found for the combined sample. Finally, the link between religion and social participation is negative and the link with economic participation is positive in Haryana. It is insignificant in Jharkhand. This shows that social cohesion is easier to attain when most of the people are Hindus. Economic cooperation tends to be more easily obtained by involving sufficient non-Hindus.

4. Conclusions and Recommendations

This Chapter presents investigations in three States where people's participation in forest management is favoured. The institutional settings are quite different. The amount of State involvement is the most distinguishing feature.

A factor analysis on indicators of participation results in two important considerations for participation: social and economic. The attitude towards meetings in the village (social participation) is the most important consideration in each case.

An econometric analysis shows under which conditions a person is most likely to choose a high level of participation. When the condition of the forest is good and/or when people are dependent on the forest, participation goes up. Low average levels of education in the family and high levels of education of the respondent enhance participation. Greater involvement of women in the community stimulates participation.

A high level of people's participation facilitates the initiation of a participatory institution. Once an institution is created, a lower level of participation is needed to keep the participatory process going. This Chapter has presented an empirical study of people's participation in forest management in various situations. Identification and recognition of a key-role of the people in forest management is a possible way to prevent deforestation.

Based on the foregoing factor analysis and multiple regression analysis, a strategy for successful forest management can be derived. It seems reasonable to suggest that the State takes the first move because they possess resources that local people lack. Taking the first move does not mean that a top-down approach should be followed. The process should commence in those villages where participation is most likely to take place, as we concluded from the regression patterns. For instance, the best chances for voluntary participation can be found among the villagers who depend highly on the forest and perceive the quality of the forest as good. Having improved the chance of a successful start for forest management, the successful village can then serve as an example for other villages to extend the process. Motivated by successes in the first stages, resources can be mobilised to replicate the process in villages with less favourable circumstances. Hence, the process should not be bottom-up either, but it should be an interaction between the State and the people, leading to a win-win situation. Transparency of the State and legal rights for the people are important aspects for success as well.

In order to improve forest management practices in India, people should be given more freedom to act on their own. The State should provide resources and assistance by formally allowing them a share in forest products. This would enhance the development of the village and the mutual trust between villagers, so that mutual participation can be sustained, by getting closer to the optimal level of social capital. As mentioned by several authors (Coleman [37]; Fukuyama [57]; Castle [29]; Woolcock [219]), it is very easy for the State to destroy social capital; it is much harder to create it. This study shows the presence of such social capital in the studied villages of rural India.

Chapter 4

Estimating a Game Theoretic Model

1. Introduction

Applied game theory generally goes as far to show possible game situations and discussing their implications.[1] This Chapter derives an estimation procedure for non-cooperative games, using cross section data. During the transformation of the data it is attempted to pick extreme points, to find the outer borders for the actual game. The estimation procedure is an algorithm that switches between several software packages. In an application a motivated choice can be made for the transformation functions such that the estimated game comes closest to the actual game of the empirical problem at hand. To demonstrate the estimation procedure, an application is made to a situation where a villager voluntarily chooses a level of participation in forest management.

Applied game theory has grown in popularity. The values of the payoffs however are generally picked illustratively. This Chapter proposes a stepwise procedure for deriving the payoffs and actions in a game from (primary) cross-section data, in order to introduce more rigor in deriving the payoff matrix.[2] Hence, this procedure starts from a diametrically different viewpoint: the game is assumed not to be known with certainty in a practical situation, but the data will tell which game is actually being played. A common economist's assumption is relaxed here to enable such an estimation. I make the assumption that the society is not necessarily in equilibrium. This gets closer to the reality than mainstream economics on which Coase comments:

> What is studied is a system which lives in the minds of economists but not on earth. (Coase [36])

This Chapter considers a game theoretic model of a situation where members of a particular organisation are free to choose whether they want to participate or not in the activities of this organisation. Examples of such organisations are a fishery community, a society for forest management, a firm, etc. An estimation procedure is derived for the game theoretic model. This Chapter focuses on the technical method to estimate a game. The difficulty in estimating such a game is to find the payoffs of actions which are not directly observable,

[1] A previous version of this chapter can be found in Lise [113]

[2] In the literature we cannot find many approaches towards estimating games. A few notable ones are Bresnahan [22], Genesove [63], and Laffont [105], which focus mainly on auction data.

but which could have been taken. To this purpose an algorithm is proposed which switches between several software packages to ultimately estimate the game, namely an econometric package like SPSS and a package handling spreadsheets like EXCEL. Observations on members across a number of organisations, about their payoffs and actions are required in the form of cross-section data. An important premise is that we do not know exactly in advance which game these members of the organisation are playing. Whether they are in equilibrium or not is not known either.

This Chapter is organised as follows. The choice situation of people's participation in an organisation is formalised in Section 2.. A taxonomy of two person symmetric games is derived to enlist all possible choice situations. Section 3. proposes the estimation procedure for a non-cooperative game. Section 3.1. deals with a method to measure the basic ingredients of a noncooperative game: payoffs and actions. Indicators of participation are used to derive the action of the interviewed person, the *challenger*, and the joint action of *other* interviewed persons in the same organisation, the *contender* (Section 3.2.). Section 3.3. presents two methods for deriving the payoff matrix: the Euclidean Cluster Method and the Homogeneous Grouping Method. Section 4. applies the estimation procedure. Data are used from Haryana. Section 5. discusses the possibility of using similar techniques for more complicated games and other situations as a scope for future research.

2. A Game Theoretic Model

It is a puzzling question why some people are more actively involved in an organisation than others. To explain differences in levels of participation, let us model such situation as a game where members of a particular organisation adopt a level of participation (=choice), while the organisation provides them additional goods (=their payoff). Let us refer to this game as the *participation game*. The simplest form of such a game consists of only two persons, who have a choice between two alternatives: to participate or not. Let us assume that both persons are identical. In general, members differ in income, power and so on, but they face the same choice situation. The assumption is plausible, as long as the differences among the members are not too large.

The participation game can be formalised as follows.

a. The set of players consists of two persons: $\{1,2\}$. Let us label the two persons as challenger i and contender $-i$.

b. Player i takes action θ_i. Abbreviate the set of actions {Participate, Defect} as {P, D}.

c. The payoff to the challenger i is $\pi_i(\theta_i, \theta_{-i})$, where θ_i is the action taken by player i. The total payoff set in this situation consists of four different numerical values: $\{x, y, a, b\}$.

In this way, the participation game is formalised as a two-person symmetric game. Table 4.1 shows the payoff-matrix. This game can easily be extended to n persons, namely by identifying one challenger and $(n-1)$ contenders. It is now more instructive to write the payoffs as the choice of the challenger $\{P, D\}$ and the number of participating contenders. For instance, in a three-person game, the third person would choose between

Table 4.1. The payoff-matrix of the participation game.

Challenger:	Contender: Participate (=P)	Defect (=D)
Participate (=P)	x, x	b, a
Defect (=D)	a, b	y, y

two matrices as shown in Table 4.1, where the payoffs consist of six numerical values: $\{P2, P1, P0, D2, D1, D0\}$. The same principle can be applied to obtain a game with more players.

A special feature of this game is that there are no restriction on the values of the payoffs. Given the derived values of the payoffs, a particular situation emerges representing either a situation of harmony or conflict. Comparison of this outcome with the actual situation in a practical situation can provide insight in the rigor of the estimation technique.

Let us assign the following important economic interpretation to a main difference in the participation game.

Definition 2..1 $x - y$ = *the net incentive to participate.*

If the net incentive to participate $x - y$ is positive, we are dealing with a meaningful game situation, where mutual participation is more beneficial than mutual defection. However, in practical situations it is not guaranteed that $x > y$. Refer to this as the *reverse* situation.

Definition 2..2 *If* $y > x$*, the net incentive to participate is negative, we are dealing with a* reverse *game.*

It is useful to classify all possible games. For that, we have to consider all 24 (that is 4!) possibilities for ordering the payoffs a, b, x, y. These can be divided broadly into six distinct cases. Let us concentrate on strict inequalities, and define games with equal payoffs as *transition games*.

In 12 out of 24 possible cases, either to participate or to defect is a dominating action.

Definition 2..3 *An action is* dominating *when it is the best response to all possible actions of the opponent.*

Then the game has a unique Nash equilibrium, which is the combination of the two dominating actions.

Definition 2..4 *An action* $\hat{\theta}$ *is said to be a Nash equilibrium for the participation game, if the following inequalities are satisfied.*

$$\pi_i(\hat{\theta}) \geq \pi_i(\theta_i, \hat{\theta}_{-i}) \text{ for all } i \in \{1, 2\} \text{ and for all } \theta_i$$

In 10 of the 12 cases the individual best response matches with the collective preferred outcome. Let us call this the *Pareto game*. In the remaining two cases, the individual best response conflicts with the collective preferred outcome. This is the well-known prisoner's dilemma.

Table 4.2. Classification of games based on the payoff ordering.

Name of the game[a]	Payoff ordering standard	reverse
prisoner's dilemma (1[b])	$a > x > y > b$	$b > y > x > a$
Pareto game[c] (5)	$x > \max\{a,y\}; b > y$	$y > \max\{b,x\}; a > x$
assurance game (1)	$x > a > y > b$	$y > b > x > a$
coordination game (2)	$x > y > \max\{a,b\}$	$y > x > \max\{a,b\}$
chicken's game (1)	$a > x > b > y$	$b > y > a > x$
battle-of-sexes game[d] (2)	$\min\{a,b\} > x > y$	$\min\{a,b\} > y > x$

[a] The prisoner's dilemma, assurance game, coordination game and the chicken's game are well-known. Two other games, Pareto game and the battle-of-sexes game are added to complete the classification.

[b] The number of possibilities is denoted in the brackets.

[c] In the Pareto game the players have a dominating strategy which coincides with the collective preferred outcome.

[d] The battle-of-sexes game is opposite to the coordination game: the Nash equilibria consist of a participating and a deviating player.

In the other 12 cases there is no dominating action, but there are always two pure-strategy Nash equilibria. In 6 of these 12 cases, mutual participation is one of the Nash equilibria of the game. Then it is purely a matter of coordination to arrive at this equilibrium. These games are known as the coordination game and the assurance game. However, in the other 6 cases, the Nash equilibria consist of a mix of participators and defectors. Then it is optimal for players to take opposite actions; to ill-coordinate their actions. These games are the well-known battle-of-sexes game and the chicken's game. Even in the battle-of-sexes game actions should be coordinated, but in such a way that players take opposite actions.

The coordination and battle-of-sexes games are very similar to, respectively, the assurance and chicken's game. There is, however, a difference. For the coordination and battle-of-sexes game the Nash equilibria payoffs are strictly greater than the remaining payoffs, which is not true for the assurance and chicken's game. Table 4.2 summarises the derived complete taxonomy of two-person two-action symmetric games.

3. Estimation Procedure

3.1. Quantifying Payoffs and Actions

3.1.1. The Amount of Incremental Income

Quantifying the amount of incremental income to a member of an organisation can best be illustrated by an example. In the case of forest management goods like fuelwood, fodder, irrigation water, grazing rights, goodwill of other people, ease in collecting resources, etc. are all gains from participation. The exact gain varies from situation to situation, depending on the particular institutional structure and how benefits are distributed.

Let the amount of incremental income π_i be the total sum of the imputed value p_{ij} of the products q_{ij} which accrue to members of a particular organisation.[3]

$$\pi_i = \sum_{j=1}^{r} p_{ij}q_{ij} \tag{4.1}$$

Adding all such separate components gives the total gain from an organisation to a particular member; the payoff.

Definition 3..1 *Payoff π_i is the incremental income accruing to person i due to the presence of a local organisation.*

3.1.2. Indicators of Participation

Various situations exist where a person can decide whether to participate or not. Some institutions encourage participation, while other institutions can do the opposite.

The choice situation which is of interest here is the extent of people's participation in a particular organisation. The basic postulate is that the choice to participate or not is voluntary. For measuring participation, it is not sufficient to know whether a person is a member of a organisation or not. The choice to become a member is made only once; it does not account for changes in perception during the process. Some people can be involved very actively, while others only reap the benefits. To separate such kind of people, it is necessary to look for alternatives in quantifying participation.

Consider the following components of people's participation.[4]

i. Contributing.

ii. Sharing in the benefits.

iii. Involvement in decision-making and evaluation.

People can choose to participate or not in many aspects of an organisation. The above mentioned aspects are three such choice situations. Participation means taking part in something.

This distinction can be used to pose questions on a person's participatory attitude. Answers to these questions can, for example, be recorded at an ordinal scale, as follows:

Not at all ________ little bit ________ medium ________ quite a lot ________ very much
1 ____________ 2 ____________ 3 ____________ 4 ____________ 5

The answers to each question can be framed so that a higher number stands for more participation. This way of collecting a response from a member of an organisation is used to stipulate the actual level of participation.

Definition 3..2 *Let PI_i^j represent the reply of person i to question j as the j-th participatory indicator of person i.*

Denote vector $(\mathrm{PI}_i^1, \mathrm{PI}_i^2, \ldots, \mathrm{PI}_i^m)$ as $\mathbf{PI}_i$. In the remainder of this Chapter, bold variables stand for vectors. In our present example, a participatory indicator can take any integer value between 1 and 5: $\mathrm{PI}_i^j \in \{1,2,3,4,5\}$.

---

[3]In a more general model the transformation of p_{ij} and q_{ij} into π_i can be assumed to be non-linear as well.

[4]Refer to Pongquan [162] for a comparative discussion on the definition of people's participation.

3.2. Deriving the Level of Participation

3.2.1. Choice of the Challenger

The participatory indicators $\mathbf{PI}_i$ of Section 3.1.2. represent a set of responses to a variable, which is not directly observable: participation. Denote the level of participation for individual i as θ_i. In general, θ_i is a function of $\mathbf{PI}_i$. Let us denote this function as μ, which converts m indicators of participation into a single value:

$$\begin{aligned}&\mu : \mathcal{R}^m \rightarrow \mathcal{R};\\ &\theta_i = \mu(\mathbf{PI}_i) = \mu(\mathrm{PI}_i^1, \mathrm{PI}_i^2, \ldots, \mathrm{PI}_i^m)\end{aligned} \tag{4.2}$$

Where m is the number of relevant questions asked during the survey and $\mathcal{R}$ is the set of real numbers.

The simplest way to define the level of participation is by taking the grand total of all participatory indicators $\mathbf{PI}_i$ as defined in equation 4.3.

$$\theta_i = \sum_{j=1}^{m} \mathrm{PI}_i^j \tag{4.3}$$

Let θ_i^{ts} denote the level of participation based on the total sum of the indicators of participation.

An alternative way of defining the level of participation is by undertaking a factor analysis on the indicators of participation (Harman [78], Klein [97], Kloek and Mennes [98]). This is a method for translating a large set of variables into a few choice-variables: the principal components.

As is shown in Section 2., it is possible that the set of participatory indicators results in more than one principal component. Since the principal components are orthogonal, each principal component reflects a different choice situation for participation. Let $\theta_i^{\mathrm{pc}k}$ denote the level of participation based on the kth-principal component of the indicators of participation.

3.2.2. Choice of the Contender

Consider the following formalisation, required in order to derive the choice of the contender. Define the set of organisations as $O := \{1, 2, \ldots, o\}$ in which a particular organisation exists. Let N_j, for $j \in O$, be the set of $|N_j|$ (surveyed) members in organisation j, where $|N_j|$ is the cardinality of set N_j. Let $N := \bigcup_{j \in O} N_j$ be the collection of all observations and let $|N|$ be the total number of people surveyed in all organisations together.

Consider the following payoff-function:

$$\pi_i(\theta_i, \theta_{-i}),$$

where θ_i is the level of participation of challenger i belonging to organisation j. The collective choice of all other members θ_{-i} is an $(|N_j| - 1)$-vector representing all the individual levels of participation in organisation j, excluding person i.

$$\theta_{-i} := (\theta_1, \ldots, \theta_{i-1}, \theta_{i+1}, \ldots, \theta_{|N_j|}) \text{ where } i \in N_j \tag{4.4}$$

In the notation introduced so far, the model is very general, where the decision of a single player affects the payoff of all other players. Then the players are known to *act strategically*. To fit the $|N_j|$-person game at the organisational level into the participation game, assume that each member is comparable, irrespective of wealth, caste or creed. This assumption also implies the symmetry of the game. Irrespective of the sample size, each particular organisation can be converted into a two-person game, by introducing an opponent: the contender.

In an empirical setting, however, it seems likely that people act more impulsively than strategically. When members *act impulsively*, they observe the outcome of the actions of all other members. It is less likely that they deduct each and every actions of all other members. Contrast this with strategic action, where the actions of each individual can be observed. It seems plausible to assume that a person's choice is conditioned as if all other members had acted jointly. In this case, the other's actions are taken as a simultaneous move.

By summarising the $(|N_j|-1)$-vector into a scalar, the situation can be analysed as a two-person game. Let λ_j be the transformation function which converts vector θ_{-i} into a single value.[5]

$$\begin{aligned} &\lambda_j : \mathcal{R}^{(|N_j|-1)} \rightarrow \mathcal{R}; \\ &\vartheta_i := \lambda_j(\theta_{-i}) \text{ for all } i \in N_j \text{ and } j \in O. \end{aligned} \tag{4.5}$$

Function λ_j converts the actions of $|N_j|-1$ *other* members in organisation j into a *composite* second person. I assume here that each member of an organisation treats all others as composite second players. The composite second player is the contender who takes action ϑ_i. In this manner, the game retains the flavor of an n-person game. The payoff function can now be written as $\pi(\theta_i, \vartheta_i)$. Consider the following two functional forms of λ_j:

I Mean: $\lambda_j(\theta_{-i}) := \overline{\vartheta}_i$.

II Variance: $\lambda_j(\theta_{-i}) := \sigma_{\vartheta_i}$.

Many other functional forms for $\lambda_j(\theta_{-i})$ are possible.[6] These are of interest because of their simplicity and complementarity.

Since we are looking at the average participatory behaviour of the other persons in a certain organisation, as opposed to each person i with a participation level θ_i, it is obvious to rely only on *their* mean level of participation, by excluding the level of participation of person i. This leads to the functional form of $\lambda_j(\theta_{-i})$ as represented in equation 4.6.

$$\overline{\vartheta}_i = \frac{1}{|N_j|-1} \sum_{\{k \in N_j | i \neq k\}} \theta_k \text{ for all } i \in N_j \text{ and } j \in O. \tag{4.6}$$

When the aggregation of the level of participation of other members, $\overline{\vartheta}_i$ is based on the mean, we call it model I. This situation can be interpreted as action risk-neutral, because all

---

[5]There are no further limiting assumptions on how function λ_j should behave.

[6]The actions of the composite second player is defined analogous to portfolio selection. Markowitz [132], for instance, develops six different ways to measure risk as a way to reduce the dimensionality of a choice problem.

actions of other members are weighed equally by taking the mean. Let us refer to model I as the *mean situation*.

When we take the variance between the levels of participation of all other members of a particular organisation with respect to the action of the contender, the situation is intrinsically different. In contrast with the mean situation where all deviations are accounted for equally, the 'variance' situation is characterised by squares of the variation from the mean of the actual level of participation. In this way, people with the highest variation are accounted for much more than the people with a small variance. Positive and negative deviations are no longer cancelling each other out. Every deviation is added to the total value of the action of the contender. This leads to the functional form of $\lambda_j(\theta_{-i})$ as represented in equation 4.7.

$$\sigma_{\vartheta_i} = \frac{1}{|N_j|-1} \sum_{\{k \in N_j\}} (\theta_k - \theta_i)^2 \text{ for all } i \in N_j \text{ and } j \in O. \tag{4.7}$$

When the aggregation of the level of participation of other members, σ_{ϑ_i}, is based on the variance we call it model II. This situation can be interpreted as action risk-averse. Let us refer to model II as the *variance situation*.

It is also possible to extend the mean and variance situation to more than two persons. An n-person game can be derived from the same data, as long as $n < |N_j|$ for all $j \in O$. Let use divide the game into one challenger and $n-1$ contenders. The contenders can be subdivided into $n-2$ contenders, which choose ϑ_i directly from the observed θ's and a composite contender. The action of the composite contender is derived from the remaining observations, which set can be denoted as $M_j(\subset N_j)$. Formulas 4.6 and 4.7 change as follows for the multi-person game.

$$\overline{\vartheta}_i = \frac{1}{|M_j|} \sum_{k \in M_j} \theta_k \text{ for all } i \in N_j \setminus M_j \text{ and } j \in O. \tag{4.8}$$

$$\sigma_{\vartheta_i} = \frac{1}{|M_j|} \Sigma_{k \in M_j} \left(\theta_k - \frac{1}{|N_j|-|M_j|} \Sigma_{l \in N_j \setminus M_j} \theta_l\right)^2 \text{ for all } i \in N_j \setminus M_j \text{ and } j \in O. \tag{4.9}$$

Here the action of the composite contender is the average or 'variance' of the level of participation of $|M_j|$ the other members in the organisation.

3.3. Two Grouping Methods

3.3.1. Euclidean Cluster Method

This Section proposes two methods on how to derive the payoff-matrix by dividing the observations into 4 groups. In the first method, we need normalised levels of participation of both the challenger and the contender in order to eliminate a bias either towards the contender or the challenger.

In general, variable s_i can be converted into a fraction between 0 and 1 by applying equation 4.10.

$$\tilde{s}_i = \frac{s_i - \min_{k \in N} s_k}{\max_{k \in N} s_k - \min_{k \in N} s_k} \text{ for all } i \in N \tag{4.10}$$

Table 4.3. Assigning members into clusters, using final cluster centers.

mean situation			variance situation		
θ	$\overline{\vartheta}$	payoff group	θ	σ_ϑ	payoff group
'large'	'large'	X (P,P)	'large'	'large'	B (P,D)
'large'	'small'	B (P,D)	'large'	'small'	X (P,P)
'small'	'large'	A (D,P)	'small'	'large'	A (D,P)
'small'	'small'	Y (D,D)	'small'	'small'	Y (D,D)

Note: $\theta, \overline{\vartheta}, \sigma_\vartheta$: levels of participation of the challenger (θ) and the contender (ϑ). P = participate, D = defect.

Where the symbol ($\breve{}$) is put above a variable to denote that it is normalised between zero and one.

The members are assigned into four clusters, so that the Euclidean distances within clusters is minimised, while the Euclidean distances between clusters is maximised.[7] Call this way of dividing the observations the Euclidean Cluster Method. The Euclidean Cluster Method also yields final cluster centers. Table 4.3 shows how to assign the final cluster centers of the actions of the challenger (θ) and the contender (ϑ) to payoff groups A, B, X, Y. Section 4. gives an example of the Euclidean Cluster Method.

Table 4.3 distinguishes between the mean and the variance situation, because they require a different interpretation. In the mean situation, a large value for the level of participation always means 'participate', while a small value means 'defect'. In the variance situation, when the variance among the actions is small, the actions of the challenger and the contender are nearly the same. However, when the variance is large, the action of the contender must be the *opposite* to the action of the challenger.

3.3.2. Homogeneous Grouping Method

Another way to derive a payoff matrix is by assigning the observations into four *equal* groups. This can be done in two steps.[8]

The first step separates participators and defectors, using the level of participation θ_i. Let us take the *median* of the level of participation θ_i as the separating threshold value θ^*.[9] All members who have a level of participation at least as large as the threshold value θ^* are assumed to participate, while others defect. In general, k members participate and $n-k$ members defect. The separation between participators and defectors can be denoted formally as follows.

$$
\begin{aligned}
P &= \{i \in N | \theta_i \geq \theta^* (\text{person } i \text{ is participating})\} \\
D &= \{i \in N | \theta_i < \theta^* (\text{person } i \text{ is defecting})\}
\end{aligned} \tag{4.11}
$$

Note that $P \cap D = \emptyset$ and $P \cup D = N$ represent all persons in the sample.

[7]This is possible through the procedure of SPSS [146].

[8]I assigned the observations into equal groups with the help of an EXCEL spreadsheet.

[9]The mean or Euclidean clusters can also be used to realise a division, but this does not guarantee an even distribution of the observations.

The second step splits these two sets into two more subsets by taking the *median* of the subset P or D as the threshold value.[10] Denote these as ϑ_P^* and ϑ_D^* respectively. In this manner, it is possible to realise four subsets of observations combining large and small levels of participation. Equation 4.12 specifies the required division.

$$\begin{array}{lll} & \text{Mean} & \text{Variance} \\ \theta_i \geq \theta^*,\ \vartheta_i \geq \vartheta_P^* \text{ (large,large)} & \Rightarrow i \in X & i \in A \\ \theta_i \geq \theta^*,\ \vartheta_i < \vartheta_P^* \text{ (large,small)} & \Rightarrow i \in A & i \in X \\ \theta_i < \theta^*,\ \vartheta_i \geq \vartheta_D^* \text{ (small,large)} & \Rightarrow i \in B & i \in B \\ \theta_i < \theta^*,\ \vartheta_i < \vartheta_D^* \text{ (small,small)} & \Rightarrow i \in Y & i \in Y \end{array} \tag{4.12}$$

Consult Table 4.3 for assigning the observations into payoff groups. Call this way of dividing the observations the Homogeneous Grouping Method.

3.3.3. Payoff Matrix

The payoff-matrix of Table 4.1 on page 47 can be computed by taking average payoffs of the assigned observations to each cluster. Formally, the payoff can be calculated by applying the following formula, where $|A|$ denotes the number of observations in payoff group A:

$$a = \frac{1}{|A|}\sum_{i\in A}\pi_i; b = \frac{1}{|B|}\sum_{i\in B}\pi_i; x = \frac{1}{|X|}\sum_{i\in X}\pi_i; y = \frac{1}{|Y|}\sum_{i\in Y}\pi_i \tag{4.13}$$

It is also possible to assign the observations to payoff groups in the k-person game ($k < n$). The k challengers and $n-k+1$ contenders can be assigned to P and D via Equation 4.11. The final composite contender can be assigned to A, B, X, Y by comparing the average level of participation of the challenger and $n-k+1$ contenders with the level of participation of the final contender via Equation 4.12. Similarly to Equation 4.13, the average payoffs can be calculated for the k-person game, by taking the averages within each payoff group.

3.3.4. Regression

As before, the payoff-matrix (Table 4.1) can be derived by calculating the average payoffs of all observations in each payoff group. Further, it is also possible to base the payoffs on several socio-economic variables, within each payoff group. This can be done by estimating the following regression equations:

$$\pi_i = \beta^T \mathbf{SE}_i + \alpha + \varepsilon; \quad i \in A, B, X \text{ or } Y \tag{4.14}$$

Where **SE** is a vector of socio-economic variables, β and α the regression coefficients, ε the error term. This results four times in a regression on subsets of the total data set using the division of equation 4.12.

Estimation of Equation 4.14 for each subset of data proceeds in two steps. Step one derives which socio-economic variables should be included in the regression equation 4.14.[11]

[10] This step is required, because high and low ϑ's are not necessarily evenly distributed over P and D.

[11] This can be done, for example, with the stepwise-regression method of SPSS [146]. This procedure selects one by one the most significant socio-economic variables as defined by the user. Only those variables with a user-defined significance level are added into the regression.

Step two regresses the final equation, which can be used for further analysis. In an alternative method relevant socio-economic variables can be selected irrespective of the resulting significance level. Such a method, although less reliable, is tailor-made for analysing the impact of certain policies.

3.3.5. Comparison

The Euclidean Cluster Method gathers the observations around the equilibrium of the game. In some cases, the equilibria of the game can be found even without calculating the average payoffs. This simplifies the analysis considerably.

In the Euclidean Cluster Method, four clusters must be calculated. This means that at least one observation is assigned to each group. It sometimes happens that the final group centers are not statistically different.[12] Most of the time, it is possible to calculate four different groups meaningfully.

The Euclidean Cluster Method is hard to use for a regression in each cluster, because the number of observations can be very unevenly distributed. The extreme possibility exists of having one observation in a cluster. This single observation solely determines the final payoff. This problem can be avoided by forcing the observations into certain groups. The Homogeneous Grouping Method is designed for that purpose.

The Homogeneous Grouping Method forces the observations into four groups of the same size. The Homogeneous Grouping Method focuses on the relative differences in the payoffs. It does not consider how close the actions of players are together. This makes the Homogeneous Grouping Method less useful for interpreting the game at the institutional level. This is why the Euclidean Cluster Method is preferred to the Homogeneous Grouping Method from a game theoretic point of view. For an analysis at the organisational level, the Homogeneous Grouping Method is the best we can do from a statistical point of view.[13]

Figure 4.1 summarises the algorithm which this Section derived for estimating a game theoretic model.

4. An Example of Estimating Payoff Functions

This Section shows how to estimate a payoff matrix in a practical situation. Data from Haryana are used in the examples for applying the Euclidean Cluster Method and the Homogeneous Grouping Method.

In order to apply the algorithm, we need to choose appropriate functions for μ and λ_j. Let the action of the challenger be based on the total sum of the participatory indicators (μ) and the action of the contender be based on the mean level of participation of all other interviewed people in a village (λ_j). Table 4.4 shows how the payoffs are assigned to a particular cluster by considering the final cluster centers. The largest cluster centers are underlined to show the meaning of 'large' and 'small' in Table 4.3. For instance, the combination of the two largest cluster centers corresponds to payoff group X. The other clusters are assigned

[12]This can be tested by running the of SPSS.

[13]Application of the Euclidean Cluster Method requires many more observation for a reliable estimation at the organisational level.

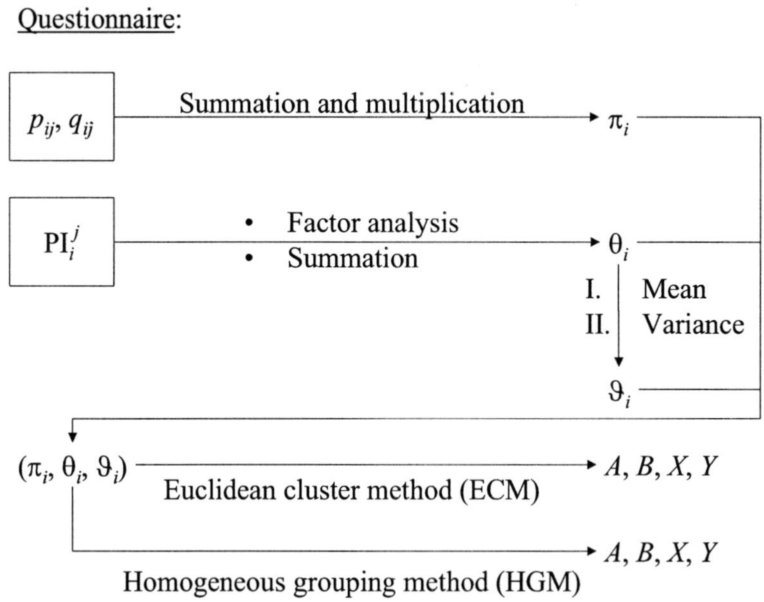

Figure 4.1. A flow chart of the estimation procedure.

to their payoff groups similarly. The number of observations in each cluster and the average payoffs are displayed as well.

The same situation is also estimated with the Homogeneous Grouping Method. Table 4.5 shows the result. The means of the payoff groups can also be compared with each other. This can be done by running the   in SPSS (Section 3.3.5.). It turns out that all differences are significant with at least 95% confidence.

As Section 3.3.4. argues, it is also possible to base the payoffs on several socio-economic variables through a regression equation, with a significance of all included variables above 95%. The values in the brackets denote the averages of the included socio-economic variables, within the payoff group.

$$x = +6.93\ (32.5)\ \text{FODTOT} + 3836\ (0.830)\ \text{FORDEP} + 0.0110\ (24606)\ \text{INCOME} + 113\ (15.0)\ \text{FWTOT} + 2.40\ (793)\ \text{BAMTOT} - 3117;$$

$$b = -0.00194\ (1529952)\ \text{CAPITAL} + 305\ (30.7)\ \text{LANQU};$$

$$a = +97.6\ (16.2)\ \text{FWTOT} + 227\ (6.23)\ \text{ANIMUN};$$

$$y = +5023\ (0.670)\ \text{FORDEP}.$$

The included variables have the meaning as shown in Table 4.6:

Let us concentrate on the total use of fuelwood, by substituting the average values of all other factors, to see the impact of changes in socio-economic variables on the payoff structure. The set of payoff-equations can be simplified as follows:

Table 4.4. An example of the Euclidean Cluster Method.

Cluster	θ^{ts}	$\overline{\vartheta}^{ts}$	π	payoff	cases
4	0.8236	0.8287	x	3766	55
2	0.6777	0.3553	b	5366	53
1	0.2804	0.6406	a	3528	8
3	0.2555	0.1533	y	1583	11

$b > x > a > y \Leftrightarrow$ Pareto game

Table 4.5. An example of the Homogeneous Grouping Method.

Group	θ^{ts}	$\overline{\vartheta}^{ts}$	π	payoff	cases
1	0.8626	0.9099	x	4151	36
2	0.7969	0.5584	b	6118	33
3	0.5070	0.6053	a	3135	29
4	0.4902	0.2020	y	3273	29

$b > x > y > a \Leftrightarrow$ Pareto game

Table 4.6. Meaning of the included variables in the example regression.

Variable	Meaning and definition
FODTOT	Total use of fodder. [quintals[a] per year]
FORDEP	Forest dependence. [fraction of the amount of goods collected from the forest and the total need of those goods]
FWTOT	Total use of fuelwood. [quintals per year]
BAMTOT	Total use of bamboo. [pieces per year]
LANQU	Total output of land. [quintals per year]
ANIMUN	Animal unit. [cow-units]

[a] One quintal is equal to 100 kg.

$x = 2466 + 113\ (15.0)$ FWTOT ;

$b = 6395;$

$a = 1414 + 97.6\ (16.2)$ FWTOT;

$y = 3365.$

A decrease in the total need for fuelwood from 15 quintals per year to 7.9, leads to a structural change in the payoff structure. The Pareto game becomes then a reverse prisoner's dilemma. Hence, whenever the payoffs become equal, due to changes in the socio-economic variables, a point of transition is reached. This leads to a structural change when the socio-economic changes are further intensified.

By estimating payoff functions, a sensitivity analysis is possible on the payoff structure and a game with varying payoffs can be derived. Variation in the values for these socio-economic variables anticipates a dynamic process.

5. Conclusions and Extensions

This Chapter has derived an algorithm for estimating a game theoretic model, based on cross-section data. The algorithm is applied to the special case of people's participation in forest management.

Possibilities to quantify two basic ingredients of a game, action and payoff, are pointed out. The level of participation is based on several participatory indicators and the payoff is based on the imputed value of the additional products from a particular organisation. The action of the challenger is taken as the grand total or as the first or second principal component of the indicators of participation. The action of the contender is based on the mean or the variance of the levels of participation of all other surveyed members of a particular organisation.

The complete data-set is divided into four groups based on the level of participation of both the challenger and the contender. This is done through the Euclidean Cluster Method or the Homogeneous Grouping Method. In the Homogeneous Grouping Method, as the name suggests, it is tried to divide the observations in groups of the same size. The payoff-matrix is derived by taking the average values of the payoffs within each group. The four payoffs of the game do not need to remain fixed, but can be based on socio-economic variables. Changes in these variables can induce changes in the relative payoff-structure.

There are possibilities to extend the model to higher dimensions. In each step of the estimation procedure, we explain how the method can be extended to a multi-person game. Extension to a repeated game is possible by simply assuming that the derived one-period game is played repeatedly. A further extension is possible by using the Homogeneous Grouping Method. By basing the payoffs on several several socio-economic variables, it is possible to formulate a repeated game with changing payoffs as well. Unfortunately, the general analytical solutions for such games are complex.

Extension of the model to a continuous choice framework seems possible with the current data set. Then we need to work with non-linear payoff functions. The first order conditions provide the Cournot reaction functions to see whether a benchmark solutions is possible as a joint utility maximisation outcome (Tirole [202]).

An area for further research is to extend this model to asymmetric games. Extension is obtained by allowing for different payoff structures π_i and π_j when $i \neq j$. For example, when person i is facing a prisoner's dilemma it is quite possible that person j is facing an assurance game, $i \neq j$. Consider, for example, the game between the State and the villagers. The priorities for the State can be quite different from the priorities of the villagers. Then they do not need to face the same game situation and the assumption on the similarity of the players is no longer valid. For instance, in the situation where the State faces a prisoner's dilemma and the villagers face an assurance game, the Nash equilibrium consists of a defecting State, while the villagers are better off by continued participation. For estimating the asymmetric game, data should first be divided into two groups, i.e. in the last example they are (1) the workers of the State and (2) the villagers. These two groups can be analysed in isolation as before to derive the game situation they are facing. A taxonomy of ten game situations for the case of asymmetric two-person games can be found in Rapoport and Guyer [167].

Chapter 5

A Game Model of People's Participation in Forest Management

1. Introduction

People have a marginal role in managing forests located in the vicinity of their villages in Northern India. This situation is scrutinised in this Chapter by studying strategic play of forest users. Thereto, a 1 versus $n - 1$ game of people's participation in forest management is estimated for three institutional and historical distinct cases at the State and village level. Critical discount factors are derived to verify whether incentives exist for villagers to mutually participate in managing commonly used forests.

The role of people in forest management in India has been low. This Chapter argues that such a marginal role is not justified, given the potential of people to participate voluntarily. Moreover, it is well-known that a higher level of people's participation may lead to a better quality of the forest, which is both beneficial for the nearby living people and the State (see for instance Poteete and Ostrom [164]). While most forests in India are under the control of the State, there also exist forests which are managed by nearby living people.

The forest councils considered in this Chapter, where the user-group and user-rights are well-defined, are a common property regime. The effectiveness of these forest councils depends, among other factors, on the amount of people's involvement. This Chapter addresses two research questions: 1) Are there incentives at the institutional level to participate in forest management? 2) Are there incentives at the village level to participate in forest management?

To analyse the incentives for participation in forest management, it is useful to distinguish, as before, between a *situation* of forest management and a forest management *dilemma* (see also Section 3.3.).

The case where forest management is a dilemma can be captured by the prisoner's dilemma (e.g. G. Hardin and Baden [75]; R. Hardin [77]; Bendor and Mookherjee [19]). In the prisoner's dilemma, private interests conflict with public interests, preventing participation to emerge as an outcome. G. Hardin [74], for instance, adopts the parable of the prisoner's dilemma to sketch the 'tragedy of the commons'.

Forest management is also problematic when it is represented by a chicken's game

(Taylor [200]). In the chicken's game, private interests again conflict with public interests, leading to an outcome of partial participation. Both the prisoner's dilemma and chicken's game constitute *collective dilemmas* (Bardhan [10, 11]; Bates [15]), as they satisfy conditions (1. and 2.) in Section 3.3. (page 25).

A situation of forest management does not have to be a dilemma. Runge [169, 170, 171], for instance, pleads that forest management is akin to the assurance game, where one prefers to do what the other does, while mutual participation has the highest preference. This is, typically, not a dilemma and it leads to a self-enforceable solution. Runge uses this metaphor to explain why 'the tragedy of the commons' has not yet emerged on a larger scale. Forest management in Runge's case exhibit effective organisations, which check-and-balance the use of forests; there is no conflict between individual and collective optimal strategies.

Basu and Mishra [14] argue that there does not need to be a dilemma in a single period in the sense of Wade [212, 213], but the dilemma shows up in the long run. This can be modeled as a repeated hawk/dove game, where people choose between the *dove* and the *hawk* strategy. Playing *dove* means sustainable forest management, while *hawk* results in forest exploitation. Playing *hawk* too often brings forests on the brink of disaster. Hence, forest degradation is not immediately apparent, but it is observed gradually over time.

Moreover, in many practical situations, people face the same choice-situation over and again. Those situations can be modelled as repeated dilemmas (Axelrod [9]; Kreps *et al.* [100, 101, 135]). A simple way to rationalise cooperation in repeated games is by considering *infinite repetitions*, which route is followed in Section 3..

While it is customary in applied game theory to identify the game being played beforehand and then continue to discuss the implications, we follow the opposite route here. The kind of strategic play in the field is verified afterwards through an estimation procedure (Chapter 4). Thereto, a field survey was conducted in rural India to collect primary data on people's willingness to participate in forest management and their benefits from such behaviour. The outcome of this estimation procedure indicates the game people have been playing at the village and State level. This result is compared with the observed situation in the field and the implications are discussed.

This Chapter is organised as follows. Rather than choosing one particular game, people's participation in forest management is modelled as a non-cooperative game, having various games as open options (Section 2.). Section 3. derives for all these options the so-called critical discount factors, which describe whether incentives exist to sustain voluntary mutual participation in forest management. Sections 4.–5. apply the model to Haryana, Jharkhand and Uttaranchal. Section 4. discusses how games at the State level are quantified and the outcomes are presented and discussed. The quantification of games at the village level along with its outcome is explained and presented in Section 5.. Section 6. provides conclusions and recommendations.

2. The Model

For studying the opportunities of local people to voluntarily participate in the management of forests adjacent to their villages, let us use a non-cooperative game model, as introduced in Section 2. of the previous Chapter. The strategy is to choose the level of participation in forest management. Here participation means the amount of involvement of a villager in a local organisation in the village to manage an adjacent forest. A participating villager adheres to rules as formulated during village meetings. This participation is awarded with the right to (partially) access the forest from which they can collect resources like fuelwood and fodder (=their net benefits or payoff), but it may also give them rights to irrigation water from erected dams in the forest or a stipend paid for protecting and maintaining the forest. Let us refer to this situation as the *participation game*.

The simplest form of such a game consists of only two persons, who have a choice between two alternatives: to participate or not. When one peasant participates, while the other does not, the single participant keeps the rules as formulated during the village meetings (following Copeland and Taylor [40]), while the other cheats on these rules, reducing the protection of the forest. This 'cheating' can be detected through 'social fencing', where the rule-abiding villager spots cheating. This can be reported to the forest guard. During a meeting of the village committee a decision can be taken on how to punish the cheater. In case of a small offence, only a warning may be issued. Finally, when both deviate, rules are not adhered to by both villagers.

Let us assume that the players of the game have the same preferences. While villagers are generally not identical, there are many situations where they are comparable; where they have equal opportunities.

Considering the motivation above, the two-person participation game can be formalised as follows.

a. The set of players consists of two imaginary entities: $\{1,2\}$. Let us label the two imaginary entities as challenger i and contender $-i$, where the challenger is an average villager, which we consider in the analysis, while the contender represents an average of other villagers, which is evaluated in the mind of the challenger. More specifically, this Chapter deals with a 1 versus $n_j - 1$ game, where n_j is the number of observations in village j.

b. Player i chooses the level of participation θ_i. While θ_i can be measured as a continuous variable, in the presentation of the results, let us focus on two choices only. Let us abbreviate the set of actions as {Participate, Defect}, or in short as {P, D}.

c. The payoff to the challenger i is $\pi_i(\theta_i, \theta_{-i})$, where θ_i is the action taken by player i. The total payoff set in the situation where two identical players have two choices, consists of four different numerical values: $\{a, b, x, y\}$.

Table 4.1 shows the payoff matrix of the symmetric two-person bi-matrix game satisfying conditions (a. to c.). For interpreting the estimated games in Sections 4.–5. use the taxonomy of two-person two-action symmetric games as presented in Table 4.2.

The robustness of mutual participation to be can be measured as the likelihood of continued mutual participation:

Definition 2..1 $\psi = \frac{x-a}{x}$ = *the incentive to participate.*

3. Critical Discount Factors

For judging the quality and stability of the organisation for forest management in the village, the participation game is repeated infinitely over time. Such a game can be analysed by considering trigger strategies, which are a code of behaviour, where the players agree to participate until a certain condition is violated, after that they defect forever to punish the deviant.[1]

For each mentioned game in Table 4.2, we derive the necessary conditions under which mutual participation can be sustained forever. The main question studied is the following: Assuming that both players participate, do they have an incentive to deviate? To answer that question, let us assign to each game a so-called *critical discount factor*, denote this as ϕ. It is well-known from the Folk theorem (Fudenberg and Maskin [55]) that such critical discount factors can be found. If the players discount their future –with discount factor δ which is equal for both players– above the critical discount factor, mutual participation can be sustained in equilibrium. The smaller the value of a critical discount factor, the greater is the likelihood to preserve mutual participation.

As discussed in the introduction, a forest management dilemma can be represented by either a prisoner's dilemma or a chicken's game. While mutual participation cannot be sustained in the finitely repeated prisoner's dilemma, it is possible with infinite repetitions, if the players follow trigger strategy σ_{it}^{pd}, assuming that they started with mutual participation.

$$\sigma_{it}^{\text{pd}} = \begin{cases} \text{P} & \text{as long as} \quad \theta_{t-1} = (\text{P,P}) \\ \text{D} & \text{if} \quad \exists i \in N, \exists \tau < t : \theta_{i\tau} = \text{D}. \end{cases} \tag{5.1}$$

This trigger strategy prescribes to participate until a player deviates. After that both players defect forever. This leads to the following critical discount factor:[2] $\phi^{\text{pd}} = \frac{a-x}{a-y}$.

The chicken's game has two pure Nash equilibria: (P,D) and (D,P). Consider the following *complex* trigger strategy:

$$\sigma_{it}^{\text{chg}} = \begin{cases} \text{P} & \text{as long as} \quad (1)\ \theta_{t-1} = (\text{P,P}) \text{ or } (\text{D,D}); \text{ or:} \\ & \qquad (2)\ \exists \tau < t : \\ & \qquad (\forall s < \tau : \theta_s = (\text{P,P}) \text{ or } (\text{D,D})) \\ & \qquad \wedge (\theta_{i\tau} = \text{D} \wedge \theta_{-i\tau} = \text{P}) \\ \text{D} & \text{if} \quad \exists \tau < t : \\ & \qquad (\forall s < \tau : \theta_s = (\text{P,P}) \text{ or } (\text{D,D})) \\ & \qquad \wedge (\theta_{i\tau} = \text{P} \wedge \theta_{-i\tau} = \text{D}) \end{cases} \tag{5.2}$$

Trigger strategy σ_{it}^{chg} prescribes to *keep on participating as long as the other did not deviate first*. Trigger strategy σ_{it}^{chg} differs from σ_{it}^{pd}, because the deviant returns to participation, when the other person implements the infinite sequence of deviations. In this

[1]For analysing repeated games, subgame perfect equilibria generally need to be found. However, in our specific case, it suffices to look at Nash equilibria, as they coincides with subgame perfect equilibria. Infinitely repeated games with trigger strategies can be analysed as supergames, which are equivalent to normal form games.

[2]This trigger strategy can be found by rewriting the equation where the payoff of always participating ($x\sum_{i=0}^{\infty}\delta^i$) is equal to the payoff of defecting ($a + y\sum_{i=1}^{\infty}\delta^i$) (see for instance Stahl [198]).

manner, the deviant can ascertain a higher payoff b instead of y. When both players follow this behaviour, then the critical discount factor can be derived similarly as in the case of the prisoner's dilemma, yielding the following critical discount factor: $\phi^{\text{chg}} = \frac{a-x}{a-b}$.

The following trivial strategy sustains mutual participation, since it is a Nash equilibrium as $x > a$.

$$\sigma_{it}^{x>a} = \{ \text{ P} \quad \text{irrespective of the choice of the other player} \tag{5.3}$$

Strategy $\sigma_{it}^{x>a}$ is not a trigger strategy, because mutual participation can be sustained through participating. Then, there is no incentive to defect ($\phi^{x>a} = 0$).

It is also possible to consider the trivial strategy which sustains mutual defection, which is again a Nash equilibrium. Superscript 'rpg' stands for *reverse Pareto game*:

$$\sigma_{it}^{\text{rpg}} = \{ \text{ D} \quad \text{irrespective of the choice of the other player} \tag{5.4}$$

This is the opposite of the previous strategy, and it recommends to never participate ($\phi^{\text{rpg}} = 1$).

Finally, it is also possible that the Nash equilibrium is a mix of participating and defecting players. In that case a player wants to take an action opposite to the action of the other player. Let us refer to these cases as 'other games'. The strategy σ_{it}^{og} can be described as follows:

$$\sigma_{it}^{\text{og}} = \begin{cases} \text{P} & \text{as long as } \theta_{-i,t-1} = \text{D} \\ \text{D} & \text{if } \theta_{-i,t-1} = \text{P} \end{cases} \tag{5.5}$$

In this case, mutual participation cannot be sustained too ($\phi^{\text{og}} = 1$).

The question, whether it is profitable to deviate from mutual participation in the past, can now be answered by summarising the critical discount factors for every possible game. The result of this Section is summarised in the following lemma.

Lemma 3..1 *If overall trigger strategy σ_{it}^{ag} in Equation 5.6, combined with Equations 5.1–5.5, is the code of behavior, then mutual participation can be sustained if and only if $\delta \geq \phi$, where critical discount factor ϕ is as given in Table 5.1.*

$$\sigma_{it}^{\text{ag}} = \begin{cases} \sigma_{it}^{\text{pd}} & \text{for the prisoner's dilemma} \\ \sigma_{it}^{\text{chg}} & \text{for the chicken's game} \\ \sigma_{it}^{x>a} & \text{for games with (P,P) a Nash equilibrium} \\ \sigma_{it}^{\text{rpg}} & \text{for the reverse Pareto game} \\ \sigma_{it}^{\text{og}} & \text{for all other games in Table 4.2} \end{cases} \tag{5.6}$$

4. Games at the State Level

In order to quantify the participation game, data are collected in Haryana, Jharkhand and Uttaranchal. As explained in Chapter 1, these cases were selected because of their differences with respect to forest quality, kind of forest resources and the way the village council is organised. At the same time these three cases all have in common that they are driven by voluntary people's participation.

Table 5.1. Critical discount factors.

Name of the game	standard	reverse
prisoner's dilemma	$\frac{a-x}{a-y}$	0
Pareto game	0	1
assurance game	0	0
coordination game	0	0
chicken's game	$\frac{a-x}{a-b}$	1
battle-of-sexes game	1	1

To study the incentives for participation, the participation game of Section 2. is quantified as follows. The payoff (π_i) is taken as the total sum of the amount of resources which are collected by the members of a village council for forest management multiplied by the average local market value (also derived via the field survey) of these resources.[3] The cost of collecting resources is assumed zero here. This is most probably reasonable for rural India, where alternative labor opportunities -competing with resource collection in the communal forests- are rare. Hence, payoffs are net benefits. More specifically, the payoff in Haryana includes besides fuelwood, fodder, fibre grass, and bamboo also the additional net incomes due to irrigation by damwater. The payoff in Jharkhand consists of the stipend paid to villagers working in the pool and the yield from the common pool. In Uttaranchal the payoff is composed of the market value of collected fuelwood, fodder and timber. There are great differences in the average payoffs, as the average payoff in Jharkhand is about two times the average payoff in Haryana, while the average payoff in Uttaranchal is about one third of the average payoff in Haryana (see Table B.8 at page 191).

The level of participation (θ_i) of the challenger is constructed in two steps. First, indicators of participation are derived by interviews with members of a village council for forest management. Each member responded to a number of questions about their contribution to, benefits from and involvement in forest management. Their psychological responses are recorded on an integer scale varying from 1 to 5, where a higher number represents a higher willingness to participate. Table 5.2 shows the average responses to each question. Inspection of this Table shows that the indicators of participation obtained the highest response in Jharkhand, in-between in Haryana and the lowest in Uttaranchal.

As explained in Chapter 3, one way to aggregate such indicators of participation into a level of participation is by performing a factor analysis on these indicators. This is possible by running the procedure of SPSS. Table 5.2 presents the first factor for the three cases considered in this study.[4] In the case of Haryana, Jharkhand and Uttaranchal the first factor respectively explains 45%, 36% and 35% of the variation.

[3]Prices differ across States. While they are almost the same in Haryana and Uttaranchal, they are the half in Jharkhand. However, the salaries in Jharkhand are also half of those in Haryana. In that sense the 'stakes' of free-riding and participating are comparable across States. And the argument of Hotte *et al* [84], that prices affect the incentive to monitor and enforce property rights, cannot be used to explain the different outcomes across States.

[4]Actually, 2, 3, 4 principal components were found in respectively Haryana, Jharkhand and Uttaranchal (see Chapter 3).

Table 5.2. First principal component based on indicators of participation and their average values.

Indicators of participation:	Principal components: Haryana	Jharkhand	Uttaranchal	Average values: Haryana	Jharkhand	Uttaranchal
Planting in the forest	0.383		0.048	2.19		2.24
Contribution to the forest/pool	0.226	0.172	0.177	3.06	3.22	2.76
Benefiting from the forest/pool	0.094	0.098	0.233	3.98	3.51	3.19
Ability to use the pool		0.082			3.67	
Benefits from using the pool		0.156			3.42	
Importance of meetings		**0.535**	0.095		4.63	3.98
Agreement with decisions	0.049	**0.832**	**0.771**	3.79	4.73	4.04
Attendance of meetings	**0.797**	**0.787**	**0.594**	4.11	4.50	3.78
Ability to influence decisions	**0.682**	**0.868**	**0.810**	4.15	4.55	3.61
Frequency of meetings	**0.792**	-0.020	0.275	3.80	4.30	2.73
Interest in the meetings	**0.640**	0.292	**0.842**	3.98	4.50	3.56
Gain from meetings	**0.611**	0.271	**0.815**	4.02	4.52	3.67
Suggesting in meetings	**0.584**	0.280	0.488	3.06	3.52	3.26
Percentage of variance explained	45.1%	36.0%	35.4%			
Number of observations	127	123	135			

Note: Numbers in bold face denote a dominating indicator (factor loading >= 0.5). An empty cell means that the information on that indicator of participation was not collected.

Within a particular village, the level of participation of the contender (ϑ_i) can be calculated by aggregating the level of participation of the *other* villagers. For aggregating the level of participation of other villagers, consider the following formalisation. Define the set of villages as $V := \{1,2,\ldots,v\}$ in which a forest council exists. Let N_j, for $j \in V$, be the set of $|N_j|$ (surveyed) members in village j, where $|N_j|$ is the cardinality of set N_j. Let $N := \bigcup_{j \in V} N_j$ be the collection of all observations and let $|N|$ be the total number of people surveyed within a State.

Following Chapter 4, the first way of aggregating the level of participation of the other villagers is by taking the mean, where a player can be considered as being insensitive towards the strategy of other players (mean situation, model I).

$$\overline{\vartheta}_i = \frac{1}{|N_j|-1} \sum_{k \in N_j \setminus \{i\}} \theta_k \text{ for all } i \in N_j \text{ and } j \in V. \tag{5.7}$$

The second way of aggregating the level of participation of the other villagers is by taking the 'variance', where a player can be considered as being *sensitive* towards the strategy of other players, as opposing decisions lead to higher values in this case (variance situation, model II).

$$\sigma_{\vartheta_i} = \frac{1}{|N_j|-1} \sum_{k \in N_j} (\theta_k - \theta_i)^2 \text{ for all } i \in N_j \text{ and } j \in V. \tag{5.8}$$

Hence, we have the triplet $(\pi_i, \theta_i, \vartheta_i)$ for all 385 interviewed households across 32 villages in three Indian States. A game can be estimated by normalising the levels of participation (θ_i, ϑ_i) of all interviews between 0 and 1 (following Equation 4.10 at page 52). The interviews can then be divided into four *payoff groups* $\{A,B,X,Y\}$ based on these normalised levels of participation. One way to do so is by the procedure of SPSS, which calculates the cluster centers. Then 'large' cluster centers correspond to 'P' and 'small' cluster centers to 'D'. This way of creating four payoff groups is called the *Euclidean Cluster Method*.

Table 5.3. The estimated games at the institutional level in three States.

Model	a	b	x	y	Payoff order	Name of the game	ϕ	ψ
The Euclidean Cluster Method:								
I.har	1969 (0)	7361 (8491)	5080 (2584)	5207 (4914)	$b>y>x>a$	Reverse prisoner's dilemma	0	0.612
I.jha	7772 (17405)	7001 (7104)	13225 (21035)	1129 (71)	$x>a>b>y$	Pareto game	0	0.412
I.utt	1849 (2012)	0 (0)	2394 (2336)	0 (0)	$x>a>b=y$	Pareto/assurance game	0	0.228
II.har	7514 (6083)	0 (0)	7139 (7973)	4530 (3341)	$a>x>y>b$	Prisoner's dilemma	0.125	-0.052
II.jha	1078 (0)	0 (0)	11240 (18718)	1797 (2173)	$x>y>a>b$	Coordination game	0	0.904
II.utt	144 (169)	0 (0)	2410 (2326)	1019 (1491)	$x>y>a>b$	Coordination game	0	0.940
The Homogeneous Grouping Method:								
I.har	5185 (3452)	9112 (12614)	5217 (2463)	6837 (5258)	$b>y>x>a$	Reverse prisoner's dilemma	0	0.006
I.jha	11973 (22591)	11963 (19558)	13688 (20776)	5923 (5645)	$x>a>b>y$	Pareto game	0	0.125
I.utt	1896 (2226)	2728 (2754)	2037 (2166)	2253 (1952)	$b>y>x>a$	Reverse prisoner's dilemma	0	0.070
II.har	5085 (3651)	8902 (12466)	5305 (2447)	6940 (5081)	$b>y>x>a$	Reverse prisoner's dilemma	0	0.041
II.jha	4841 (6091)	6921 (5558)	18730 (26697)	13292 (22430)	$x>a>b>y$	Coordination game	0	0.742
II.utt	1433 (1699)	1959 (2732)	2806 (2164)	2730 (2265)	$b>y>x>a$	Coordination game	0	0.489

Note: I = the mean situation; II = the variance situation; har = Haryana, jha = Jharkhand, utt=Uttaranchal; ϕ = the critical discount factor; The number in the brackets denotes the variance in the payoff within the payoff group.

Alternatively, the sample could also be split into four payoff groups of an almost equal size, with the help of an EXCEL spreadsheet. This can be done in two steps. The first step separates participators and defectors, using the level of participation θ_i. Let us take the *median* of the level of participation θ_i as the separating threshold value θ^*. The separation between participators and defectors can be denoted formally as shown in equation 4.11.

The second step splits these two sets into two more subsets by taking the *median* of the subset P or D as the threshold value. Denote these as ϑ_P^* and ϑ_D^* respectively. In this manner, it is possible to realise four subsets of observations combining large and small levels of participation. Equation 4.12 specifies the required division. This way of creating payoff groups is called the *Homogeneous Grouping Method*. Finally, the payoffs can be calculated by applying the formula of equation 4.13.

Table 5.3 shows the games at the State level, which are estimated using the Euclidean Cluster Method and the Homogeneous Grouping Method. Furthermore, a distinction has been made concerning the sensitivity of the challenger towards the strategy of the contender (mean and variance situation). Hence, there are four possible outcomes per State. The Euclidean Cluster Method leads to the most reliable result from a game theoretic point of view, as the method clusters together villagers with similar levels of participation. Hence, the first 6 games of Table 5.3 are most likely to represent the actual game being played. There are, however, still two options. Out of these two options, the mean situation seems to provide the best representation of the reality, as the estimation results for the Euclidean Cluster Method and the Homogeneous Grouping Method are the closest. This implies that the consulted villagers seem to be *insensitive* towards differences with the levels of participation of other villagers.

In the case of the Euclidean Cluster Method and the variance situation, which is quite different from the mean situation, no participation is found in Haryana in the short term (prisoner's dilemma), while conditional participation is found in Jharkhand and Uttaranchal (coordination game). Table 5.3 indicates that the estimation result is quite robust, as the incentive to participate ψ is over 0.90 in Jharkhand and Uttaranchal. In Haryana a value of $\psi = -0.05$ is found, which indicates that the prisoner's dilemma may not be a very severe conflict. Moreover, the critical discount factor $\phi = 0.125$ is rather low and mutual participation is still a likely outcome in the long run.

In the case of the Euclidean Cluster Method and the mean situation, mutual participation can be sustained in all three cases. In Haryana a higher payoff is possible by mutual defection (reverse prisoner's dilemma), while unconditional participation is found in Jharkhand (Pareto game) and a somewhat more conditional participation is found in Uttaranchal (Pareto/assurance game). This typically represents the actual situation. Participation is difficult to achieve in Haryana, because of unequal landholding and a great difference between rich and poor. In Jharkhand the villagers are willing to work in the pool, as there are hardly any good labor alternatives. In Uttaranchal the people in the village are quite equitable and are willing to participate when others are willing too.

Table 5.3 shows for the case of the Euclidean Cluster Method and the mean situation, a higher incentive to participate than in the variance situation. This indicates that, if a villager is insensitive towards the level of participation of other villagers, mutual participation is more likely to emerge. Table 5.3 shows that the difference between x and a is smaller. The likelihood to deviate from mutual participation (ψ) is 0.61 for Haryana, 0.41 for Jharkhand and 0.23 for Uttaranchal, which means that the organisation for forest management is less stable once villagers are indifferent to the level of participation of other villagers.

Table 5.4 shows the average level of participation (final cluster centers) in each payoff group.[5] High levels of participation are underlined to distinguish between high and low levels of participation, which are labeled as 'Participate' and 'Defect' in Table 4.1. Table 5.4 shows that the number of observations is quite unevenly distributed. While the amount of mutual participation in the mean situation in Haryana is around 27% (payoff group '*PP*'), the amount of mutual participation in the mean situation in Jharkhand and Uttaranchal varies from 63% to 82% respectively. In the variance situation, mutual participation is much more common and varies from 74% in Haryana to 97% in Jharkhand. This clustering of observations into payoff group X indicates a high number of participating villagers. In addition, Table 5.4 has no observation in payoff group B in variance situation in Haryana and Jharkhand, which demonstrates a limitation of the Euclidean Cluster Method. In this case, let us take the payoff equal to zero, as the action-pair (P,D) is not observed.

It is also helpful to plot the observations on the strategies of the challengers and the contenders graphically. Figures 5.1–5.6 presents the observations and the division of the observations according to the Euclidean Cluster Method (by using different symbols for the observations) and the Homogeneous Grouping Method (by using one vertical and two horizontal lines, representing the median values). Inspection of Figures 5.1–5.6 shows that the division of observations according to the Euclidean Cluster Method is based on relatedness, while the division of observations according to the Homogeneous Grouping Method is

[5]It was necessary to consider 5 instead of the usual 4 clusters in the mean situation in Haryana, otherwise the difference between some final cluster centers would not be statistically significant.

Table 5.4. Final cluster centres and the number of households within the clusters.

Model I.har: reverse prisoner's dilemma

Cluster	θ	$\bar{\vartheta}$	π	cases
1	0.21	0.32	*y*	9
2	0.67	0.54	*x*	34
3	0.55	0.10	*b*	48
4	0.04	1.00	*a*	1
5	0.84	0.39	*b*	35

Model II.har: prisoner's dilemma

Cluster	θ	σ_ϑ	π	cases
1	0.40	0.08	*y*	29
2	0.05	0.60	*a*	4
3	0.64	0.03	*x*	56
4	0.85	0.09	*x*	38

Model I.jha: Pareto game

Cluster	θ	$\bar{\vartheta}$	π	cases
1	0.11	0.30	*y*	2
2	0.87	0.63	*x*	77
3	0.63	0.69	*a*	20
4	0.84	0.13	*b*	24

Model II.jha: Coordination game

Cluster	θ	σ_ϑ	π	cases
1	0.00	1.00	*a*	1
2	0.67	0.07	*x*	25
3	0.88	0.04	*x*	94
4	0.31	0.44	*y*	3

Model I.utt: Pareto/assurance game

Cluster	θ	$\bar{\vartheta}$	π	cases
1	0.74	0.84	*x*	111
2	0.88	0.00	*b*	1
3	0.38	0.85	*a*	19
4	0.17	0.29	*y*	4

Model II.utt: Coordination game

Cluster	θ	σ_ϑ	π	cases
1	0.29	0.21	*y*	11
2	0.88	0.80	*b*	1
3	0.06	0.81	*a*	3
4	0.73	0.06	*x*	120

Notes: θ is the level of participation of the challenger; $\bar{\vartheta}$ is the level of participation of the contender based on the mean; σ_ϑ is the level of participation of the contender based on the variance; har = Haryana, jha = Jharkhand, utt=Uttaranchal; $\{a, b, x, y\} \in \pi$ is the payoff or net benefit from participation.

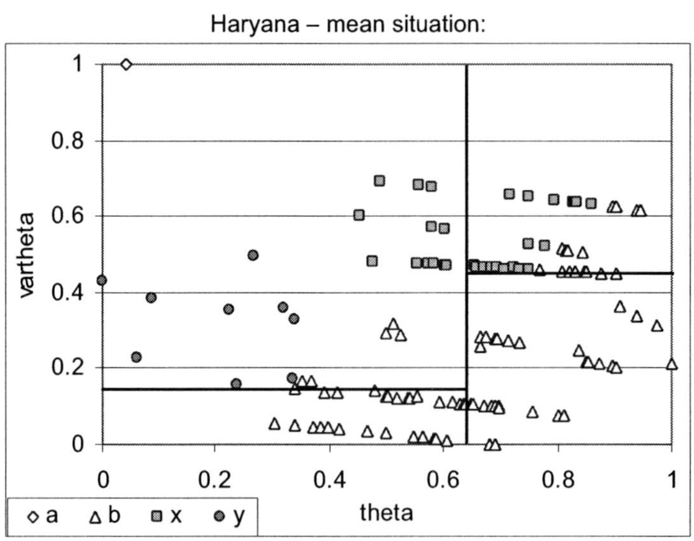

Figure 5.1. Scatter plot of strategies in the mean situation in Haryana.

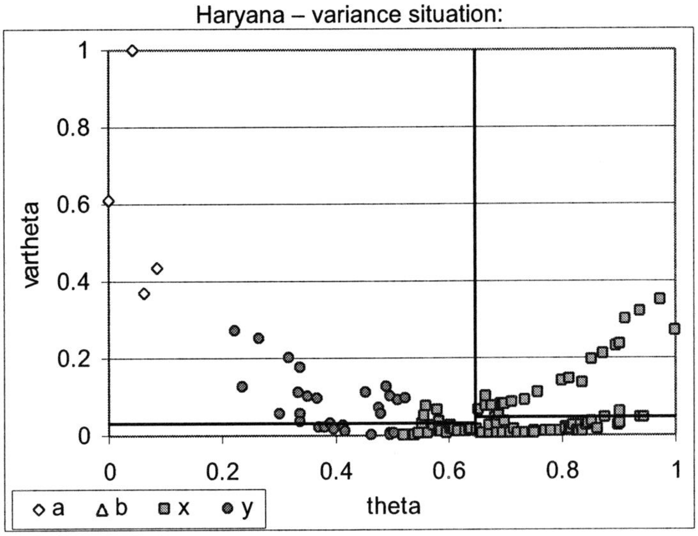

Figure 5.2. Scatter plot of strategies in the variance situation in Haryana.

more arbitrary. While the division in Figure 5.1 is quite different for the Euclidean Cluster Method (based on five clusters), as compared to the Homogeneous Grouping Method, the estimated game is still the same reverse prisoner's dilemma.

Inspection of Figures 5.1–5.6 indicates that the division based on Euclidean Cluster Method and Homogeneous Grouping Method do differ. When the average payoffs of these quite distinct divisions lead to a similar dilemma, as in the case of the mean situation, then it is most likely that we have found the actual game being played.

5. Games at the Village Level

The previous Section addressed the first research question, namely whether there are incentives at the institutional level for villagers to participate voluntarily in forest management. Let us now turn to the second research question, to verify whether there are incentives at the village level.

The Homogeneous Grouping Method is used here to estimate games at the village level. Such an approach is viable as the estimated games at the State level are quite similar to the ones found with the Euclidean Cluster Method in the mean situation. For instance, Table 5.3 shows that the games in Haryana and Jharkhand are the same, while only the game in Uttaranchal changes from a Pareto/assurance game to a somewhat different reverse prisoner's dilemma.

To explore the games at the village level, the data are divided into four equally sized payoff groups via the Homogeneous Grouping Method at the village level. The average payoffs, standard deviations and the resulting games at the village level are presented in Tables 5.5–5.7.

In order to interpret the games at the village level, let us consider whether mutual partic-

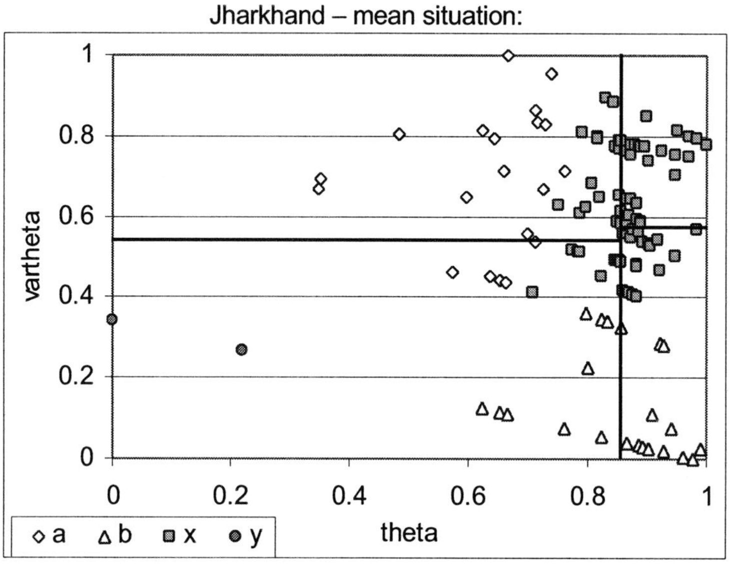

Figure 5.3. Scatter plot of strategies in the mean situation in Jharkhand.

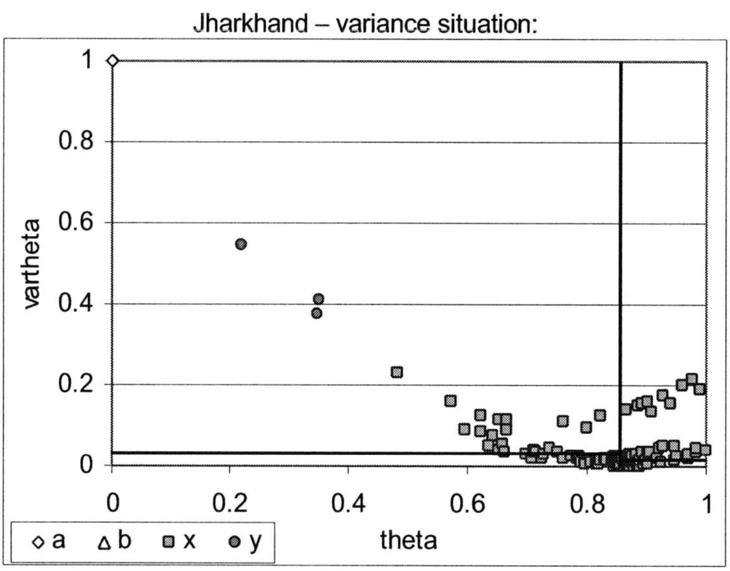

Figure 5.4. Scatter plot of strategies in the variance situation in Jharkhand.

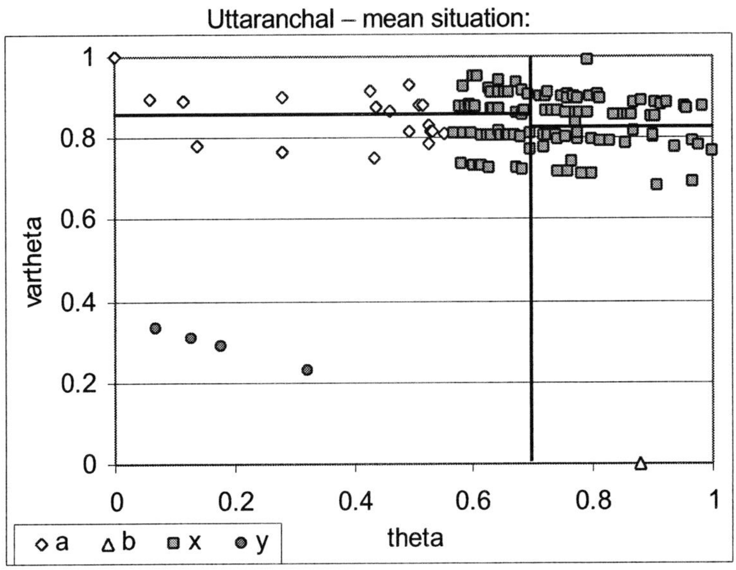

Figure 5.5. Scatter plot of strategies in the mean situation in Uttaranchal.

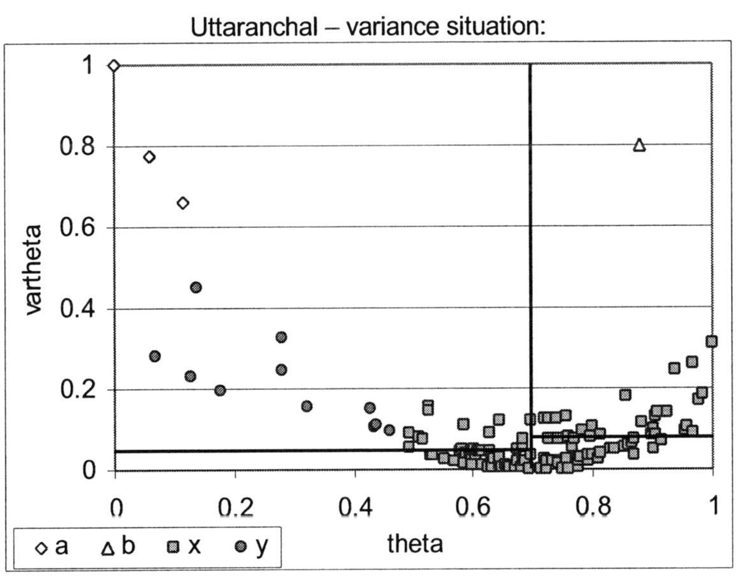

Figure 5.6. Scatter plot of strategies in the variance situation in Uttaranchal.

Table 5.5. The estimated games at the village level in Haryana.

Village name:	a	b	x	y	payoff order	Name of the game	ϕ	ψ
Panchkula district:								
Sukhomajri (19)	2712 (1917)	7003 (2217)	11448 (8325)	6636 (5763)	$x > b > y > a$	Pareto game	0	0.763
Dhamala (16)	3079 (2384)	6606 (4730)	12497 (8220)	5661 (4594)	$x > b > y > a$	Pareto game	0	0.754
Main Nada (11)	5029 (4282)	10457 (2610)	7505 (2429)	7776 (294)	$b > y > x > a$	Reverse prisoner's dilemma	0	0.330
Harijan Nada (4)	1969 (0)	4605 (0)	5183 (0)	4193 (0)	$x > b > y > a$	Pareto game	0	0.620
Godam (30)	3963 (1136)	4717 (1254)	4012 (348)	4004 (1739)	$b > x > y > a$	Pareto game	0	0.012
Ambala district:								
Mandpa (11)	9518 (6342)	14207 (4251)	32543 (35561)	3860 (1003)	$x > b > a > y$	Pareto game	0	0.708
Masoompur (12)	3266 (2207)	3865 (371)	3208 (1144)	6260 (2517)	$y > b > a > x$	Reverse Pareto game	1	-0.018
Thathar (14)	6181 (3326)	5168 (2950)	4708 (2048)	5913 (3869)	$a > y > b > x$	Reverse Pareto game	1	-0.313
Yamuna Nagar district:								
Salehpur (10)	6541 (4991)	4593 (555)	8589 (2813)	9648 (1135)	$y > x > a > b$	Reverse coordination game	0	0.238

Note: The values in brackets denote the standard errors. The choice of the challenger (= level of participation) is based on the first principal component of the indicators of participation. The choice of the contender is based on the mean of the levels of participation of other villagers. The number in the brackets in the first column denotes the number of observations per village, while the number in the brackets of the payoffs denotes the variance in the payoff within the payoff group.

Table 5.6. The estimated games at the village level in Jharkhand.

Village name:	a	b	x	y	payoff order	Name of the game	ϕ	ψ
Barwadih subdivision:								
Hendehas (7)	6596 (3239)	21338 (0)	6719 (9251)	46388 (0)	$y > b > x > a$	Reverse assurance game	0	0.018
Chapri (16)	713 (539)	5068 (5578)	4551 (5257)	5111 (5639)	$y > b > x > a$	Reverse assurance game	0	0.843
Barhania (16)	6608 (4454)	8889 (6239)	9511 (1423)	9762 (579)	$y > x > b > a$	Reverse coordination game	0	0.305
Sindhorwa (9)	39154 (55373)	82415 (5807)	52206 (45211)	78309 (0)	$b > y > x > a$	Reverse prisoner's dilemma	0	0.250
Muru (6)	840 (1188)	12361 (0)	12663 (12462)	24291 (0)	$y > x > b > a$	Reverse coordination game	0	0.934
Daltenganj subdivision:								
Bhusaria (9)	3906 (5523)	0 (0)	18226 (16152)	7811 (0)	$x > y > a > b$	Coordination game	0	0.786
Bakhari (9)	108 (153)	0 (0)	5220 (6429)	1991 (2816)	$x > y > a > b$	Coordination game	0	0.979
Kumbhawa (7)	4818 (433)	4390 (91)	4663 (391)	4470 (0)	$a > x > y > b$	Prisoner's dilemma	0.446	-0.033
Sakanpirhi/ Tandwa (16)	8194 (9971)	3219 (5021)	15442 (11940)	7989 (8678)	$x > a > y > b$	Assurance game	0	0.469
Mundaria (10)	4044 (5996)	13412 (1489)	3284 (5688)	5084 (6744)	$b > y > a > x$	Reverse chicken's game	1	-0.231
Kashia (9)	4345 (6144)	1159 (1639)	6443 (3183)	4141 (5856)	$x > a > y > b$	Assurance game	0	0.326
Khamdih (9)	4489 (5697)	5087 (5498)	8714 (182)	4559 (5862)	$x > b > y > a$	Pareto game	0	0.485

Table 5.7. The estimated games at the village level in Uttaranchal.

Village name:	a	b	x	y	payoff order	Name of the game	ϕ	ψ
Pauri district in the Garhwal region:								
Ulli	896	1013	711	1211	$x>a>b>y$	Reverse Pareto game	1	-0.259
(14)	(1122)	(887)	(1388)	(708)				
Than	428	252	335	335	$a>x=y>b$	Prisoner's dilemma/ Reverse Pareto game	1	-0.280
(8)	(80)	(37)	(473)	(473)				
Thamana	0	0	0	0	$a=b=x=y$	Game without gain	1	
(5)	(0)	(0)	(0)	(0)				
Nishni	3724	4675	3252	3150	$b>a>x>y$	Battle-of-sexes game	1	-0.145
(23)	(2215)	(1840)	(2186)	(3711)				
Rudraprayag district in the Garhwal region:								
Bachher	127	1640	936	927	$b>x>y>a$	Pareto game	0	0.864
(13)	(177)	(922)	(1070)	(758)				
Makkumath	787	1277	1326	1290	$x>y>b>a$	Coordination game	0	0.406
(15)	(493)	(1127)	(777)	(1725)				
Ushara	2080	4219	3373	2781	$b>x>y>a$	Pareto game	0	0.383
(6)	(610)	(0)	(3917)	(0)				
Sagar	1936	2906	4349	50	$x>b>a>y$	Pareto game	0	0.555
(5)	(0)	(0)	(983)	(0)				
Bageshwar district in the Kumaun region:								
Shama	3012	3777	3358	4056	$y>b>x>a$	Reverse assurance game	0	0.103
(22)	(2096)	(6061)	(1728)	(1904)				
Panyali	5625	3515	5287	3238	$a>x>b>y$	Chicken's game	0.160	-0.064
(8)	(1325)	(79)	(1647)	(1409)				
Reema	1216	1714	1390	2150	$y>b>x>a$	Reverse assurance game	0	0.125
(16)	(858)	(2568)	(1041)	(1652)				

ipation can be sustained or not when the game is played repeatedly. As argued in Section 3., when the critical discount factor (ϕ) is zero, mutual participation can always be sustained. Mutual participation cannot be sustained when the critical discount factor is one. A fraction between zero and unity denotes the non-trivial discount factor. While we find one prisoner's dilemma and one chicken's game, the results show that mutual participation can be sustained in most of the cases. Here, the data are used as a guide in deciding upon the likelihood of mutual participation to be sustainable. Another indicator of importance is the incentive to participate (ψ), which shows the likelihood of continued mutual participation.

The outcome in Tables 5.5 to 5.7 can also be compared with experimental laboratory studies with common-pool resource games. A common outcome of such experiments is that, at the aggregate level, subgame perfect equilibria organise the data quite well, while, at the individual level, subjects rarely follow Nash behaviour. Moreover, experiments shows that repetition, communication and threats increase the likelihood of continued mutual participation (see for instance Keser and Gardner [96]; Kagel and Roth [91]; Ostrom *et al* [153]; Walker *et al* [214]). The household data, we are using here, come from villages where the level of participation is set in the context of repeated interaction, where communication is possible and punishments can be expected via the village meetings. These circumstances may explain why we only find two CPR *dilemmas* out of 32 cases.

Table 5.5 shows high net benefits from mutual participation in Sukhomajri, Dhamala and Mandpa. These villages all have functional irrigation dams. The failure of access to damwater shows up in Masoompur and Thathar, where mutual participation is associated with the lowest possible payoff. Main Nada could compensate its loss of damwater by a collective purchase of a diesel pump. The single caste village Harijan Nada also has incentives for participation. In Godam, where villagers weave baskets from bamboo collected from the communal forest, mutual participation is fragile, as the incentive to participate $\psi = 0.01$ only.

It is reassuring to find a critical discount factor of 0 in Sukhomajri (Mishra [138]) and

Salehpur (Lise [114]), as these villages are known as the most successful cases in Haryana. In Sukhomajri the participatory process once started, where villagers and State representatives came to an agreement over sharing the benefits. Salehpur is known as the more recent successful revival of Sukhomajri. However, at the time of the interviews, the perception of the villagers in Salehpur was not positive, which may explain why the Nash equilibrium of mutual participation is less beneficial than the Nash equilibrium of mutual defection.

Table 5.6 shows that the estimated games have a regional linkage in Jharkhand. In the Barwadih subdivision, we only find reverse games, where mutual participation always can be sustained ($\phi = 0$). In the Daltenganj subdivision, we find in most villages that $x > y$, while the result is more mixed. In the most remote village Kumbhawa, which is about 40 kms away from the headquarters in Daltenganj city, we find a prisoner's dilemma with a critical discount factor $\phi = 0.45$, while the incentive to participate $\psi = -0.03$ only. In Mundaria we observe only partial participation. Hendehas, Barhania, Bakhari, Kashia and Khamdih are the other 'new' villages (planting effort since 1994). There mutual participation is unlikely to break down, as planting in the pool is still going on and funding of labor in the pool is still taking place.

Among the 'old' villages (planting effort before 1994), Muru is the only village where mutual participation has already broken down, mainly due to destruction of the pool by wildlife. Table 5.6 shows indeed a reverse coordination game, while mutual participation is still a Nash equilibrium. In the other 'old' villages, mutual participation is still sustainable, but it is not sure whether this will be the case in the future. Mutual participation is under pressure in Bhusaria, because of difficulties in protecting the pool. In Tandwa & Sakanpirhi, the people are losing faith in the pooling process, as the common pool did not yet substantially benefit them. Finally, the pooling process is benefiting only a part of the people in Chapri.

Table 5.7 presents the games in villages of Uttaranchal. The location of the villages can be divided into three geographical regions, namely the Pauri, Rudraprayag and Bageshwar district. As expected there are no incentives for participation in Thamana, where the communal forest is virtually gone. In Ulli, only a part of the villagers are actually interested in preserving the forest.

It is intuitive to find the highest incentive for participation in the most remote Rudraprayag district. This district is also famous from the *chipko* movement.[6] In Sagar, where the president of the forest council is suspended, mutual participation is expected to break down, but this does not show up in the data. The meetings are proceeding well in Sagar, but conflicts emerge in the field, when they try to carry out the agreed-upon rules.

In Panyali, the people are paying a yearly fee for collecting resources from the communal forest, this may be the reason that the people are not interested in contributing to planting in the communal forest. This may also be the reason why we find a non-trivial critical discount factor of 0.16. Nevertheless, the absolute net benefits are the highest in Panyali, which also indicates a high quality of the communal forest. In Shama and Reema, where theft of forest products is common, participation is conditional, where the Nash equilibrium of mutual defection is more beneficial ($y > x$) than the Nash equilibrium of mutual participation.

[6]In a protest women embraced trees and successfully avoided large-scale tree-cutting. See also Section 2.4.

6. Conclusions and Recommendations

The main objective of this Chapter has been to use data on indicators of participation in forest management and net benefits from the forest to derive what kind of game is being played among users of communal forests. Three distinct institutional settings were chosen, because of their successes in involving people in forest management on the one hand, and the diversity in achieving this, on the other hand. While the literature suggests that the game played by forest users may be usefully modeled as either a prisoner's dilemma, a chicken's game, or an assurance game, we extend the possibility to a number of other games too. Moreover, rather than choosing a game beforehand, the data has been used to verify whether villagers act in harmony with each other or not. In addition, this choice-situation is also modeled as a repeated game where villagers voluntary choose their level of participation in forest management. Trigger strategies are formulated to calculate critical discount factors above which mutual participation can be sustained in equilibrium.

By estimating games, we have also derived a quantitative measure for the effectiveness and stability of an organisation for forest management. Moreover, we can conclude from the estimation results that the studied organisations for forest management in India vary in their effectiveness. Participation is difficult to achieve in Haryana, because of unequal land-holding and a great difference between rich and poor. In Jharkhand the villagers are willing to work in the pool, as there are hardly any good labor alternatives. In Uttaranchal the people in the village are quite equitable and are willing to participate when others are willing too. This most likely outcome is based on the assumption that villagers are insensitive to variation in the level of participation of other villagers. The studied organisations for forest management in India turn out to be more stable when villages are sensitive to variation in the level of participation of other villagers in Jharkhand and Uttaranchal. In Haryana, organisational stability can only be guaranteed in the long run, when villagers discount the future above 0.125.

We do find non-trivial critical discount factors (with a value between zero and one). Moreover, we find that mutual participation can be sustained in the majority of the cases, namely in 23 out of the 32 studied villages the critical discount factor is 0, while in two more villages we find a critical discount factor of 0.45 and 0.16. In the remaining 7 villages the critical discount factor is 1. This outcome is also comparable to experimental laboratory studies with common-pool resource games, which have shown that repetition, communication and threats increase the likelihood of continued mutual participation. As this compares well to the household data we are using here, it may explain why we only find two common-pool resource dilemmas in the studied villages.

The main conclusion from a computational point of view is that the estimated games at the village and the State level do match well with the actual situation in the field. Hence, the estimation method can be recommended for application to any situation where actors have to solve collective action problems in managing a common resources.

Chapter 6

Further Analyses of the Forest Management Game

1. Introduction

This Chapter extends the game of the previous Chapter by considering trigger strategies with a finite number of deviations.[1] One scenario, where the payoff decreases due to resource depletion under no participation, is worked out for the case with changing payoffs. The model is applied to one case of forest management to show how the stage game differs over time.

Repeated games are a useful devise for modeling practical situations. The main power of a game theoretic approach is to describe a practical situation as a metaphor. However, for more fine-tuning, we require other techniques which go beyond a simple game theoretic representation. Game theory is most powerful when it is used as a substitute in the analysis; the better the main problem at hand is understood, the more precise a game theoretic model can be targeted.

This chapter focuses on linking formal game theory to a practical situation. The model is driven by and applied to the voluntary choice of villagers to participate in forest management to make it more cost effective. The analysis of this Chapter is by no means restricted to this particular application alone. Moreover, it can be applied to any situation where actors have to solve collective action problems in managing a common-pool resource. However, in order to have a fruitful application, it is desired to study the problem at hand thoroughly to make sure that the most crucial actors are selected for the application.

An outline of the Chapter follows below. The analyses of Sections 2. and 3. are mainly based on the repeated prisoner's dilemma, but the same reasoning holds for the chicken's game as well. Section 2. shows the consequences of finite-period trigger strategies. An extension of the standard repeated game towards a repeated stage game is analysed in Section 3. where the payoffs change due to the amount of resources extracted in the past. The last section summarises the results and provides areas for future research.

---

[1] A previous more extended version of this chapter can be found in Lise [111].

2. Finite-Period Deviations

It has been long debated in economic theory as to why there is always so much discrepancy between the solution of finite and infinite horizon problems.[2] Intuitively, the finite case should converge to the infinite case. The main difficulty is that it is sometimes more difficult to get an explicit expression for the finite case than for the infinite case. This is also true when a finite number of deviations is considered in the repeated prisoner's dilemma to punish the deviant. Nevertheless, an explicit expression for the critical discount factor when the number of punishments is finite comes closer to a practical situation, where people generally do not retaliate forever to punish rule-violators.

In this section, I consider the infinitely repeated prisoner's dilemma with a trigger strategy of finite period punishments. The objective is to find an expression for the critical discount factor which can sustain mutual participation in equilibrium, when the number of punishments is variable but finite. Define n as the number of punishments. Consider the following trigger strategy, where the deviant is punished by an n-period deviation.

$$\sigma_{it}^{n} = \begin{cases} \text{participate} & \text{as long as } \theta_{t-1} = (\text{P,P}) \text{ or} \\ & \theta_{i,t-j} = \text{D}\,;\, j \in \{1,2,\ldots,n\} \\ \text{defect} & \text{otherwise.} \end{cases} \tag{6.1}$$

Assume that the contender deviates in period τ, then rewriting Inequality 6.2 derives the critical discount factor, where the penalty consists of a sequence of n deviations.

$$x\sum_{t=0}^{\tau+n}\delta^{t} > x\sum_{t=0}^{\tau-1}\delta^{t} + a\delta^{\tau} + y\sum_{t=\tau+1}^{\tau+n}\delta^{t} \tag{6.2}$$

Condition 6.2 can be rewritten as follows:

$$\delta > \frac{a-x}{a-y} + \frac{x-y}{a-y}\delta^{n+1} \tag{6.3}$$

Condition 6.3 can be rewritten as an $(n+1)$-degree polynomial.

$$\begin{array}{l} f_n(\delta) = \delta^{n+1} - (1+C)\delta + C < 0; \\ \text{where } C = \frac{a-x}{x-y} (>0) \end{array} \tag{6.4}$$

It is easy to see that $\delta = 1$ is a root of f_n for every n. Then f_n can be rewritten as follows:

$$g_n(\delta) \equiv \frac{f_n(\delta)}{\delta - 1} = \delta^{n} + \delta^{n-1} + \ldots + \delta^{2} + \delta - C = \sum_{k=1}^{n}(\delta)^{k} - C > 0 \tag{6.5}$$

Denote the solution of $g_n(\delta) = 0$ as δ_n. The proof of Lemma 2..1 shows that g_n has only one positive real root.

Lemma 2..1 *Equation $g_n(\delta)$ has a unique positive real root.*

[2] An introduction to decentralised solutions in infinite horizon economies can be found in Majumdar [130]. Classic coverages of the issues can be found in Koopmans [99], Majumdar [129], and Malinvaud [131].

Proof. It is sufficient to prove the existence and uniqueness of the positive real root of $g_n(\delta)$.
i) Existence:
If $\delta = 0$, then $g_n(\delta) = -C < 0$, because $C > 0$.
If $\delta = C$, then $g_n(\delta) = \sum_{k=2}^{n}(C)^k > 0$, because $C > 0$
Therefore, $g_n(\delta)$ must intersect the x-axis at least once in the interval $[0, C]$.
ii) Uniqueness:
It suffices to show that $g_n(\delta)$ is strictly increasing for $\delta > 0$:

$$\frac{dg_n}{d\delta} = \sum_{k=1}^{n} k\delta^{k-1} > 0, \text{ for all } \delta > 0. \quad \text{QED}$$

Since we know there exists just one real root, we need to restrict ourselves to values less or equal to one, in order to have a useful discussion about discount factors.

Definition 2..2 *The critical discount factor for an n-period punishment trigger strategy is* $\phi_n = \min\{1, \delta_n\}$.

Inequality 6.3 shows that the critical discount factor with finite punishments converges to the critical discount factor with infinite punishments, when the number of punishments goes to infinity, using $\delta \in (0,1)$.

Lemma 2..3 $\lim_{n\to\infty} \phi_n = \phi^{\text{pd}}$

Theorem 2..4 *As soon as* $\phi_n < 1$, *the critical discount factors are strictly ordered:* $\phi_n > \phi_{n+1}$.

Proof. Lemma 2..1 has already shown that $g_n(\delta)$ has one positive unique real root. The strict order can be derived by subtracting $g_{n+1}(\delta_{n+1})$ from $g_n(\delta_n)$.

$$\begin{array}{rcll} \sum_{k=1}^{n}(\delta_n)^k & = & C & \\ \sum_{k=1}^{n+1}(\delta_{n+1})^k & = & C & - \\ \hline \sum_{k=1}^{n}(\delta_n)^k - (\delta_{n+1})^k & = & (\delta_{n+1})^{n+1} & > 0 \Rightarrow \delta_n > \delta_{n+1} \quad \text{QED} \end{array}$$

The intuitive idea behind Theorem 2..4 is that whenever the number of punishments increases, the likelihood of getting mutual participation as the equilibrium outcome should increase as well, because the punishment is becoming more severe.

Cases $n = 1$ and $n = 2$ In general, an n-degree polynomial has n roots which are either real, complex and/or multiple by the fundamental theorem of algebra. To solve equation "$g_n(\delta) = 0$" is a pure mathematical problem. The general solution of a n-degree polynomial has so far not been found in mathematical theory. Let us deal with a special case of this general problem. For $n \leq 2$ an explicit solution exists.

When $n = 1$, the real positive root of $g_1(\delta)$ can be used to define the critical discount factor.

$$\phi_1 = \min\{1, \frac{a-x}{x-y}\} \tag{6.6}$$

Table 6.1. The critical discount factors and the number of punishments.

Name of the game	ϕ_1	ϕ_2	ϕ_∞
prisoner's dilemma	$\min\{1, \frac{a-x}{x-y}\}$	$\min\{1, -\frac{1}{2}+\sqrt{\frac{1}{4}+\frac{a-x}{x-y}}\}$	$\frac{a-x}{a-y}$
chicken's game	$\min\{1, \frac{a-x}{x-b}\}$	$\min\{1, -\frac{1}{2}+\sqrt{\frac{1}{4}+\frac{a-x}{x-b}}\}$	$\frac{a-x}{a-b}$

Lemma 2..5 *If $\phi_1 < 1$ or $a-x < x-y$, then mutual participation is possible when the actual discount factor, $\delta \geq \phi_1$*

Lemma 2..5 shows that mutual participation can be preserved with *the threat of only one punishment* if the discount factor is large enough. However, the critical discount factor has risen to ϕ_1 which is strictly larger than ϕ_∞. The strict inequality follows directly from Theorem 2..4.

When $n = 2$, the explicit expression for the real positive root of $g_2(\delta)$ can be used to define the critical discount factor.

$$\phi_2 = \min\left\{1, -\frac{1}{2}+\sqrt{\frac{1}{4}+\frac{a-x}{x-y}}\right\} \tag{6.7}$$

So far, the analyses in this section has been based solely on the prisoner's dilemma. The same result can be derived for the chicken's game by substituting b for y in the results. The explicit expressions for the one-period, two-period and infinite-period punishment critical discount factors are summarised in Table 6.1 for the prisoner's dilemma and the chicken's game. In all other game situations the critical discount factors are *invariant* to the number of punishments, and they stay equal to the ones as presented in Table 4.2.

3. Changing Payoffs

Repeated games are generally based on a fixed payoff structure. However, for games of practical interest, this needs not be the case. Thereupon, let us consider an infinitely repeated stage game, i.e. a game with changing payoffs. Let us find behavioral strategies which sustain mutual participation in equilibrium when the payoffs change over time. Section 3. addressed this problem for fixed payoffs, using trigger strategies as the behavioral strategy. Generalising this problem to changing payoffs is problematic in the sense that when the payoff-matrix undergoes a structural change, trigger strategies can lose their credibility.

Let us focus on changing payoffs within the payoff-order of the prisoner's dilemma.[3] Then, the following restriction is in order, where an index t has been added to the payoffs

[3]Janssen *et al.* [85] consider a scenario, where payoffs are constant when both players participate, the payoffs decline when one player participates and the other deviates, while the payoffs decline most rapidly when both players defect. They derive the number of punishments to preserve participation, assuming no discounting.

to denote changing payoffs over time, while $\{a_t, b_t, x_t, y_t\}$ have the same meaning as the values in the payoff-matrix in Table 4.1 on page 47:

Assumption 3..1 $a_t > x_t > y_t > b_t$

A credible threat can be derived by applying the trigger strategy σ_{it}^{pd} (Equation 5.1 on page 62) used earlier, as the behavioral strategy. Then, the critical discount factor is characterised by Equation 6.8 and a deviation at time τ is not beneficial if the discount factor is large enough, i.e. $\delta > \phi$.

$$\sum_{t=0}^{\infty} x_t \delta^t > \sum_{t=0}^{\tau-1} x_t \delta^t + a_\tau \delta^\tau + \sum_{t=\tau+1}^{\infty} y_t \delta^t \tag{6.8}$$

Let us construct a scenario, where the forest is preserved well, until a deviation takes place. It degrades linearly after deviation. This scenario is pointed out in the following two assumptions.

Assumption 3..2 *The forest is preserved well until a deviation takes place:* $x_t = x, y_t = y, a_t = a, b_t = b,\ t < \tau$.

Assumption 3..3 *After deviation, the forest is depleted. This reduces the payoffs. Let* $\alpha (\in (0,1))$ *be the fraction of the last payoff, then* $x_t = \alpha x_{t-1}$, $a_t = \alpha a_{t-1}$ *and* $y_t = \alpha y_{t-1}$, *for* $t > \tau$. *This can be rewritten as:*
$x_t = \alpha^{t-\tau} x, a_t = \alpha^{t-\tau} a, y_t = \alpha^{t-\tau} y, b_t = \alpha^{t-\tau} b,$ *for* $t \geq \tau$.

Under this scenario it is possible to derive an explicit expression for the critical discount factor. When all the values of the two assumptions are substituted into Inequality 6.8, the following inequality can be derived. It is assumed, without loss of generality, that a deviation takes place at time $\tau = 0$.

$$x \sum_{t=0}^{\infty} \delta^t > a + y \sum_{t=1}^{\infty} (\alpha\delta)^t \tag{6.9}$$

Rewriting Expression 6.9 gives us the following polynomial expression, keeping in mind that $(1-\alpha\delta) > 0$, since $\alpha, \delta \in (0,1)$.

$$f(\delta) = \alpha(a-y)\delta^2 - [\alpha(a-y) + (a-\alpha x)]\delta + (a-x) < 0 \tag{6.10}$$

Now it is possible to find the critical discount factor .

Theorem 3..4 $f(\delta)$, *being a parabolic, has a unique real root in the interval between zero and one. This is the critical discount factor. It has the following value:*

$$\bar{\phi} = \frac{[(1+\alpha)a - \alpha(y+x)]\delta - \sqrt{[(1+\alpha)a - \alpha(y+x)]^2 - 4\alpha(a-y)(a-x)}}{2\alpha(u-y)}$$

Proof. It must be shown that $f(\delta)$ of Condition 6.10 has only one real root in the interval $(0,1)$. For $\delta = 0$, $f(0) = a - x > 0$. For $\delta = 1$, $f(1) = (\alpha - 1)x < 0$. Since $f(\delta)$ is a parabolic, there can be only one intersection in the interval $(0,1)$, and $f(\delta)$ is negative after the intersection until 1, satisfying Condition 6.10. Because $f(\delta)$ is a U-curve and $f'(0) = \alpha(a-y) - (a - \alpha x) < 0$, the critical discount factor sought for must be the smallest root of $f(\delta) = 0$. QED

Theorem 3..5 compares the critical discount factor to the fixed-payoff situation.

Theorem 3..5 *If $y > 0$, then $\bar{\phi} < \phi^{\text{pd}}(= \frac{a-x}{a-y})$*

Proof. Rewrite $f(\delta)$ as $c\delta^2 + d\delta + e$, where:

$$
\begin{aligned}
c &= \alpha(a-y) \\
d &= -\alpha(a-y) - (a - \alpha x) \\
e &= a - x
\end{aligned}
$$

$$\frac{-d - \sqrt{d^2 - 4ce}}{2c} \overset{?}{<} \phi^{\text{pd}} \Rightarrow -e \overset{?}{>} \phi^{\text{pd}}(c\phi^{\text{pd}} + d) \overset{\text{(actual values)}}{\Leftrightarrow}$$

$$-(a-x) > \frac{(a-x)}{(a-y)}\left[\frac{(a-x)}{(a-y)}\alpha(a-y) - \alpha(a-y) - (a - \alpha x)\right] \Leftrightarrow$$

$$-(a-y) > \alpha(a-x) - \alpha(a-y) - (a - \alpha x) \Leftrightarrow 0 > (\alpha - 1)y \Leftrightarrow y > 0$$

This inequality was already assumed to hold. QED

The last theorem shows that when a deviation lowers the payoffs and the payoff under mutual defection is positive, a more severe threat is possible, making mutual participation more likely to occur.

4. Conclusions and Discussion

A relation between the critical discount factor and the number of periods of punishment is derived. It is mathematically proved that the critical discount factor is strictly decreasing in the number of punishments for deviation. This proves the intuitive idea that mutual participation is easier to preserve, when the threat consists of a longer sequence of deviations. The extension to finite-period trigger strategies is therefore refreshing and has empirical appeal. However, more work needs to be done to see what happens with other threats of potential punishments to rule violators. Further research on finite-period punishments, in the field of mathematics, is possible by considering elliptic curves to approach a general solution of an n-degree polynomial.

The impact of changing payoffs due to a change in the level of resources is worked out for one particular scenario. It is assumed that resources degrade linearly as soon as a deviation takes place, this leads to decreasing payoffs for all players. If the payoffs are positive, then a deviation is less attractive in this scenario as compared to the situation with fixed payoffs.

Another way of approaching such situations is by using so-called recursive strategies, where past deviants are punished as soon as the present payoff structure allows for that. The construction of such recursive behavioral strategies is complicated. First, the game in the last period and the present game should be common knowledge. Second, the actions of the last period need to be known as well. Using this information, a player can check whether a deviation took place in the previous period and if it is possible to punish the deviant now. The choice of the information set, which is available to the players, is very important.

Incorporating institutional change into the model is complicated. The participation game can only describe the effect of a changing institutional environment on the payoffs.

Description of the institutional environment itself is still outside the model. Yet, the level of participation is in itself an indicator of social capital and institutional performance.

Further research on games with changing payoffs is important, because the solution set differs from the solutions for repeated games with fixed payoff structures. It is not an easy task to deal with such games, because the plan of action of the players needs to be constructed.

Part III

CASE STUDIES

Chapter 7

The State as Initiator: Joint Forest Management in Haryana

1. Introduction

This Chapter presents a qualitative study about the specific situation of joint forest management in Haryana. The idea of social fencing was born after negotiations between the Haryana government officials and villagers from Sukhomarji. This idea has subsequently been replicated to other villages. However the outcome of the process of the building of the dams by Haryana government officials and protection of the forest by villagers has led to mixed outcomes. This chapter sheds light on the various types of conflicts and makes suggestions for resolving them.

It has been increasingly recognised that regeneration of forests is possible only by involving local people in its management (Chopra *et al* [31], Palit [154], Sarin [175], Poffenberger and McGean [160]). One such community oriented programme is joint forest management. Enlisting people's participation is the main challenge for successful forest management.

Regeneration of forests is only possible by involving local people in its management. This was realised in Haryana after successful experiments in Nada and Sukhomajri. It has drawn the attention of many politicians, non-governmental organisations, researchers and the government of Haryana.

This process which is known as *joint forest management* is a collaboration between the State and the people, where the government of Haryana plays a mediating role. Since it is not easy for the people on their own to regenerate a degraded forest, the villagers need to escape from the poverty trap first.[1] Thereupon, the forest department decided to provide dams, with the main purpose of arresting soil run-off, but also to enhance agricultural productivity, so that people can improve their livelihood, i.e. their life standard. The philosophy of this process is

> to link the economic interests of villagers living adjacent to forest areas with sustainable management of those areas. (SPWD [197, page 43])

[1] People are said to be caught in a poverty trap when they are unable to improve their backward position in the society on their own. See, for example, Burkey [28].

In Haryana State, through the joint forest management programme, 14148 hectares of protected forests are jointly managed by villagers and the forest department. There are 45 registered societies, known as Hill Resource Management Societies. In 36 Hill Resource Management Societies, about 65 dams have been constructed (see Table B.2 at page 186 for further details).

A brief account of the Chapter layout is given here. This Chapter starts with the evolution of people's involvement in forest management in Haryana to describe its context and basic features. The evolutionary process is divided into three phases: experimental, replication, and finally institutionalisation. The first experiments took place in Sukhomajri and Nada. This experience was replicated rapidly in the second phase. The speed of replication slowed down in the third phase. The institutionalisation phase resembles the actual joint forest management programme in Haryana (Section 2.). The present experience of joint forest management is studied village-wise, to point out the difficulties in implementing the joint forest management programme at the field level (Section 3.). In Section 4., the village-wise studies are compared with each other to provide basic features of joint forest management in Haryana. This Chapter is ended with a summary and the main conclusions for improving joint forest management in Haryana.

2. The Evolution of Joint Forest Management

This Section discusses the historical setting which ultimately led to the notification on joint forest management. In addition, I point out institutional changes in forest management in the studied northern part of Haryana. I also present the basic philosophy of the joint forest management programme by defining its objectives.

2.1. Phase One: Evolution of the 'Sukhomajri Model'

More than 80 years ago the villagers were the 'owners' of the forest, according to the treasurer who is residing in Dhamala village. More specifically, the forests were an open access resource, as there was no management whatsoever. Such a situation is typically prone to mismanagement (see Section 3.). It ultimately led to heavy degradation of these forests. Subsequently, the government took over the forest next to the villages and labelled them as protected. Since then, the Shivalik hills, foothills of the Himalayas, have been under the control of the State. The State faced much difficulty in regenerating these now degraded forests which once used to be dense. It was very difficult for the State to stop people from entering these forests. Moreover, the new plantations of the State were destroyed repeatedly.

In 1958 a dam, costing Rs[2] 1 *crore*,[3] was built in Chandigarh, the State capital, after the design of Le Corbusier, the initiator and designer of the union territory of Chandigarh. This dam resulted in an artificial lake known as the Sukhna lake. However, 16 years after the creation of that lake, its water storage capacity had been reduced by 60 percent, due to accumulation of silt coming from the Shivalik hills. The cost of removing sand, silt and clay that accumulated was estimated to be more than Rs 10 *crore* (Mishra and Sarin [139]).

[2]In 1996, 1 US dollar was equal to 36 Indian rupees (Rs).

[3]One *crore* is equal to ten million.

Sediments were fed into the lake during the yearly monsoon rains. To solve the problem of sedimentation, officials from the Central Soil and Water Conservation Research and Training Institute, under the leadership of Mr P. R. Mishra, a forest officer, walked through the river 'Kansal Choe' and inspected the surrounding forests in 1974. At the root of the river there was a village called Sukhomajri, where they found goats grazing freely in the nearby Shivalik hills.[4] Fertile soil was getting washed away from the hill-slopes by the yearly monsoon rains. Grazing was identified as a major cause for degradation of the hills and, hence, the siltation of Sukhna lake. During the survey it was found that out of 85 hectares of forest, adjacent to Sukhomajri, only 7 ha were fit for planting with trees.

However, the people of Sukhomajri were not interested in preserving Sukhna lake, which is only used for recreation and only benefits the rich of Chandigarh city. Daulat Ram, a resident from Sukhomajri, came with the proposal to construct a dam in the back of the village. He posed the following question to the officials of the Central Soil and Water Conservation Research and Training Institute:

> Why do you want us to protect your trees to save your Sukhna lake? Our problems have so far always been neglected. You have plenty of money to protect the Sukhna lake. Why don't you give us something that will benefit the people of our village as well. (Mishra [138])

Thereupon, the first earthen dam was erected in Sukhomajri, based on the design of Dr Jodi Keller in 1976. Water from the dam was distributed with sprinklers to irrigate crops. A catchment area[5] of 4.3 hectares was demarcated, while the command area[6] was 6 hectares (Dhar [43]). The land that was irrigated with water from the dam showed a great increase in output of cultivated food-grains.

In return for building the first dam, Mr P. R. Mishra proposed that the people should ban free grazing in the forest. Until and unless they stall-feed their cattle, it would be impossible to regenerate the forest. In return, the people are getting irrigation water from the earthen dam. The fodder needs should be met through other means, like manual collection. To equally distribute the dam water, it was decided during a meeting between the villagers that everybody is given an equal water right, irrespective of the amount of land owned. In this way, landless farmers could also obtain a share of the profit by leasing their water rights to landowners.

After the first dam, three more dams were constructed. The second dam is most successful and it does not even dry up during the summer. Its command area is 16–24 ha.[7] Two other dams initially functioned well, but now have a very small capacity, due to siltation, and they are no longer used for irrigation. The first dam is now mainly in use for washing the buffaloes and letting them drink.

[4] A more detailed account of the evolution of the Sukhomajri model as well as the experience in Nada can be found in Mishra [138] and Mishra and Sarin [139]

[5] The catchment area is the total part of the hills that contributes water to the artificially created pond.

[6] A command area is the size of land that can be irrigated by the dam.

[7] The exact size of the command area depends on the amount of rainfall during the monsoon. After a good monsoon it can irrigate up to 28 ha of land.

2.1.1. Second Experiment in Nada

Before replicating this people's oriented way of forest management on a larger scale, it was decided to undertake one more experiment in Nada. Sukhomajri is a unique village, where land-ownership is quite equal and most of the people belong to the *Gujjar* caste. These equitable conditions are not prevalent in Nada. Only if the process could work in a situation as in Nada as well, there would be enough confidence for replicating it on a larger scale.

The relatively large village Nada consists of four hamlets, namely Harijan Nada, Upper Nada, Middle Nada and Lower Nada. Two separate societies were created to avoid caste related problems: one in Harijan Nada and the other in the remaining three hamlets: Main Nada. In Main Nada most of the people belong to the high *Lawana* (=*Rajput* or fighter's) caste. In Harijan Nada, as the name suggests, most people belong to the lowest *Harijan* caste, which is the focus of the remainder of this Section.[8]

According to the continuously reelected president Mr Pyare Lal the situation of Harijan Nada was essentially different from Sukhomajri. The villagers of Harijan Nada had their right to the hills from the very beginning, while the people of Sukhomajri were illicitly grazing their goats and cattle. That is why the people of Sukhomajri could be persuaded more easily to stall-feed their animals. The existing law had to be reenforced only. This was more difficult in Harijan Nada. Initially, the trees which were planted in the hills were demonstratively cut by the people. In response, many meetings were held to convince the people that they would ultimately share in these trees as a reward for their protection. Today, the hills are again covered with trees.

The lease of *bhabbar*[9] became a good source of income for the Hill Resource Management Society in Harijan Nada. In addition, two dams have been constructed in Harijan Nada. One of these dams provided enough water to irrigate all the private lands of the *Harijans*. About 14 families own land, while their Hill Resource Management Society consists of 20 members. One *Rajput* family also owns land in the command area of the water-providing dam, but they were not living among the *Harijans*. As one of the rules of the Hill Resource Management Societies only residents of the village can get water. In order to get access to this water, this family constructed a house and started to live in Harijan Nada.

2.2. Phase Two: Replication of the Sukhomajri Model

Since the experiences in Nada were quite satisfactory, the Haryana forest department decided to replicate the Sukhomajri model on a larger scale.[10]

In the second phase, funds created by the Haryana forest department, were used to select sites for the construction of dams. Several dams were built from these funds. It was also decided that in some villages, people could start to lease *bhabbar* grass, as their reward for protecting the forest. Both the dams and the *bhabbar* leases enhanced the economic position of quite a number of people. The number of registered Hill Resource Management Societies

[8]For a classification of castes see footnote 4 on page 18.

[9]The botanical name for *bhabbar* is *eulaliopsis binata* (Varalakshmi [210]). It is also known as elephant grass. It is a strong fibre grass and it is used for rope-making and paper.

[10]Mr P. R. Mishra continued to develop this model in Jharkhand. This is known as *chakriya vikas pranali*: a system of cyclic development. This is discussed in Chapter 8.

was rapidly increasing. However, the importance of institution-building at the village level was not yet realised. This became the focus of the third phase.

2.3. Phase Three: Institutionalising Joint Forest Management

Institution building at the village level, through the joint forest management programme, consists of three phases.

1. Initiation: This should result in a joint management agreement.
2. Initial management period.
3. Ongoing implementation.

Prior to joint forest management, the interests of the people have to be matched with the interests of the forest department through negotiations, to obtain a successful collaboration between the two parties. This important bargaining process between the forest department and the villagers in Haryana has been dealt with extensively by Chopra *et al.* [31, Chapter 6]. They formalised this into a Nash bargaining game. Below, I discuss the interests of the people on the one hand and the State on the other hand. The negotiation game between these two parties can be typically framed into a prisoner's dilemma. For instance, in the case where villagers and the State can choose between cooperating or not cooperating with each other, the dominating strategy of both the villagers and the State is typically not to cooperate. However, both parties are better off by mutual cooperation: the creation of a win-win situation. When the negotiation process leads to a contract, which is known as a joint management agreement, and when the prisoner's dilemma is overcome, a solid basis is laid for undertaking joint forest management.

2.3.1. Mutual Benefits from Joint Forest Management

The Haryana forest department is interested in and responsible for sustainable management of the hills. To achieve a sustainable situation, the hills which are most degraded need to be replanted first. Once the forest cover of the hills are beginning to take a good shape again, the forest department can increase their revenue share from the forested hills. Key to regenerating the forest is the banning of free grazing in areas of new plantations and preferably in the whole forest. Under ideal circumstances, institutions are created at the village level where the villagers closely check each other to protect the forest. This is also known as *social fencing*. The Haryana forest department obtains the following benefits for the State when it participates in joint forest management:

1. Lower cost of forest management and an increased revenue from the forest in the long run.
2. Environmental restoration.
3. Improved relation with the villagers.

Yet, villagers generally only become interested in protecting the forest once they can secure a clear benefit from it. Hence, there is a need to *link the forest to the economic interests of the villagers*, because as long as the food security is not guaranteed, the people are less likely to participate in protection of the forest and replanted new forests on barren land. Moreover, sustainable forest management needs to acknowledge the stakes of the villagers. The incentive list for the villagers for participating in joint forest management consists of one or more of the following benefits for the villagers.

Water: The agricultural production can be augmented by irrigation water through a channel system which is linked to a dam.

Fish: The dams which contain water throughout the year can also be leased for fish farming.

Fodder: Villagers can obtain the lease-right for fodder from the State. In general people are allowed to collect fodder after paying a nominal fee, known as *dati*, to the treasurer.

Bhabbar: The villagers can obtain the lease-right for collecting *bhabbar* from the State.

Bamboo: Those communities whom practice the traditional craft of basket-making, can obtain individual permits for bamboo. The fees are Rs 7 per 100 pieces of bamboo. The collection permit is given in the period from October until June.

Timber, firewood, etc.: As a reward for protecting the new plantations and the forest the people can also share in long-term products. Initially, it has been proposed that the share of villagers should be respectively 50% when the forest was badly degraded or 25% when the forest was well stocked before the initiation of joint forest management (Sarin [176]). However, the joint forest management notification of Haryana is not clear on the rights of the community on timber:

> In the case of produce like timber, fruit, medicinal herbs, etc. Haryana forest department's net income from sales *could* be shared with Hill Resource Management Societies, which may invest some of its share of the income back into the joint management area for increasing its productivity further. (Arora and Khare [7, page 47])

All benefits involve a certain cost to give the people a sense of their value, otherwise the people take them as valueless which may encourage depletion of the resources.

2.3.2. Contributions of The Energy and Resources Institute

The joint forest management monitoring programme of The Energy and Resources Institute has been funded by the Ford Foundation. They are the mediating non-governmental organisation, acting as a catalyst between the forest department and the villagers. They have strong links with both actors. They are the communicating body through which the problems of the people could be heard and responded to. The classical hierarchical way of managing the forest by the forest department (see Figure 7.1) has been altered by the introduction of Working Groups at two levels (see Figure 7.2).

The bottom-up approach was introduced in the following ways.

1. Creation of environmental awareness at the village level through village meetings.
2. Creation of awareness of people's problems by the forest department. People's involvement in forest management should have the main priority.
3. Conducting research on various aspects of joint forest management to improve the implementation of the programme.
4. Organising workshops to train several people from the forest department and the villages. (Vijh and Arora [211])

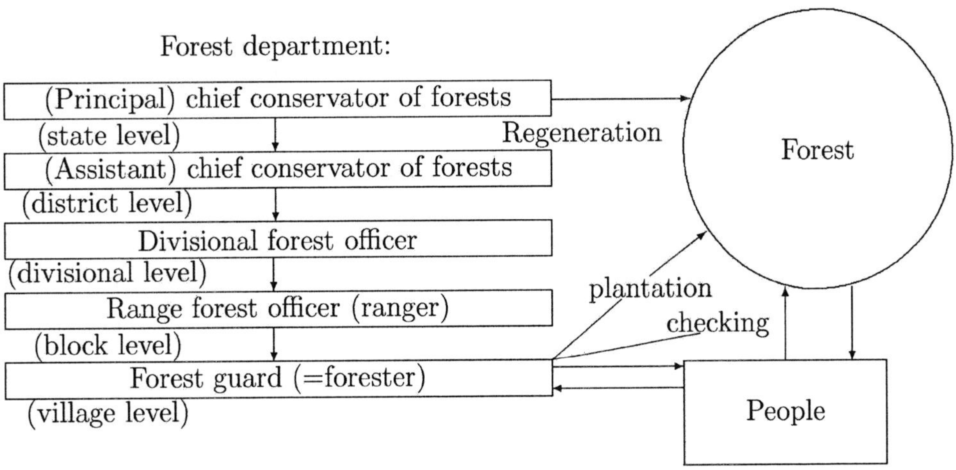

Figure 7.1. Typical hierarchical management of forests.

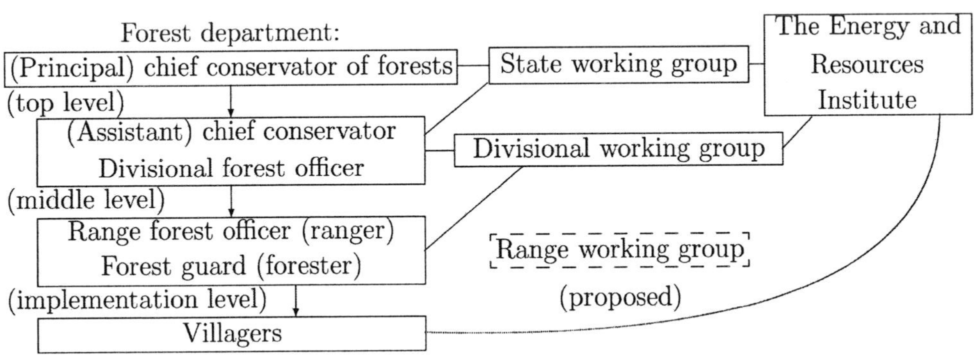

Figure 7.2. Bottom-up management of forests.

Sources: HFD and TERI [79], Vijh and Arora [211].

2.3.3. The Evolution of Hill Resource Management Societies

As soon as a joint management agreement is reached, the village institution can become officially registered as a Hill Resource Management Society (HRMS). This Hill Resource Management Society has an executive committee consisting of a president, a treasurer, a secretary and a number of managing members. They are elected from and by the villagers, which are members of the Hill Resource Management Society. The elected management team manages other common village resources as well, as these are related issues. The following rules and responsibilities exist for Hill Resource Management Societies (HFD and TERI [80, translated from Hindi]).

1. They should plan for protecting the forest.
    a. Stop free grazing.
    b. Appoint a guard among the villagers to stop free grazing.
2. Rules for establishment of a society.
    a. Every family is a member. The forest department needs signatures from both men and women to form a society.
    b. The managing body should have 7 to 11 members. All castes should be equally represented. At least two women should be in the managing body.
    c. At least 3 general body meetings should be held per year.
    d. Accounts should be audited every year.
    e. It is compulsory to keep a bank account.
3. Rules for managing the dams.
    a. Proper planning should be done to maintain the dams.
    b. Water from the dam should be equally distributed.
    c. The catchment area of the dams should be protected.
4. Rules for leasing *bhabbar*.
    a. If desired, the society can get the *bhabbar* leased from their forest.
    b. Preference of the *bhabbar* lease is given to societies with dams.
    c. The lease can be purchased for the average yield of the last three years.
    d. Other villages should also share in the *bhabbar* yield to the extent that it is not sold outside the village.
    e. If the *bhabbar* cannot be used privately it can be leased out with the help of the forest department.
    f. The *bhabbar* should be cut by the villagers.
5. Rules for leasing fodder grass.
    a. Poorer people should get concession on the cost of the grass.
    b. Grass should go to those people who need it most.
    c. The implementation of these rules should be discussed in meetings.

Money collected from the members of the Hill Resource Management Society like sales of fodder, *bhabbar*, fish and water is put in the village fund. The village fund is used for the development of the village and for maintaining the dam. The forest department would also like to see that the income of Hill Resource Management Societies is reinvested into regeneration of the forest.

So far, 45 Hill Resource Management Societies have been registered in Haryana. The growth in the total number of Hill Resource Management Societies is shown in Figure 7.3.

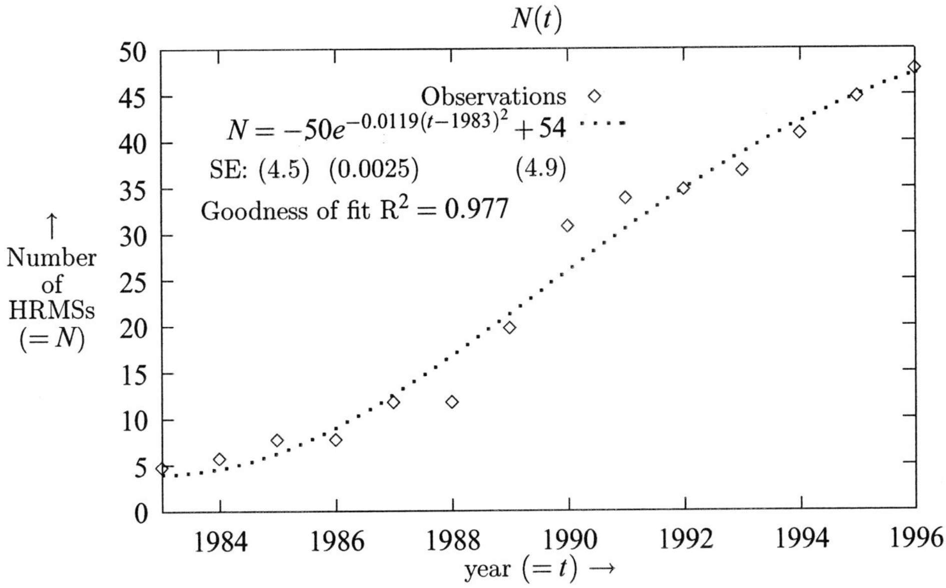

Figure 7.3. Evolution of Hill Resource Management Societies over the years.

Note: Standard errors (SE) are denoted in the brackets.

Its growth process is typically an *S*-curve. The fit, with non-linear regression, has a goodness of fit of 97.7%. Initially, the growth was zero: the experimental phase [1974–1982], because no Hill Resource Management Society was yet registered. The growth was very fast when the forest department was solely in charge: the replication phase [1983–1990]. The growth rate became more moderate when The Energy and Resources Institute came in: the institutional phase [1991–1996]. Their objective was not to maximise the number of Hill Resource Management Societies, but to improve the functioning of them, so that other villages follow this example in the longer run.

3. Village-Wise Case Studies

To see how the principles of Section 2. are practised in the field, an impression is given by discussing the performance of joint forest management in a number of villages as covered during the fieldwork. Figure 7.4 presents a map of the location of the fieldwork undertaken in Haryana. Figures 7.5 and 7.6 provide two illustrations of the situation in the field, namely two succesful villages: Sukhomajri and Salchpur.

3.1. The Panchkula District

3.1.1. The Present Welfare State of Sukhomajri

Two local institutions co-exist in Sukhomajri village, where about 90 families reside. In addition to having a Hill Resource Management Society, Sukhomajri also belongs to a *panchayat* (village council), which is jointly with three other villages, namely Vasdevpura,

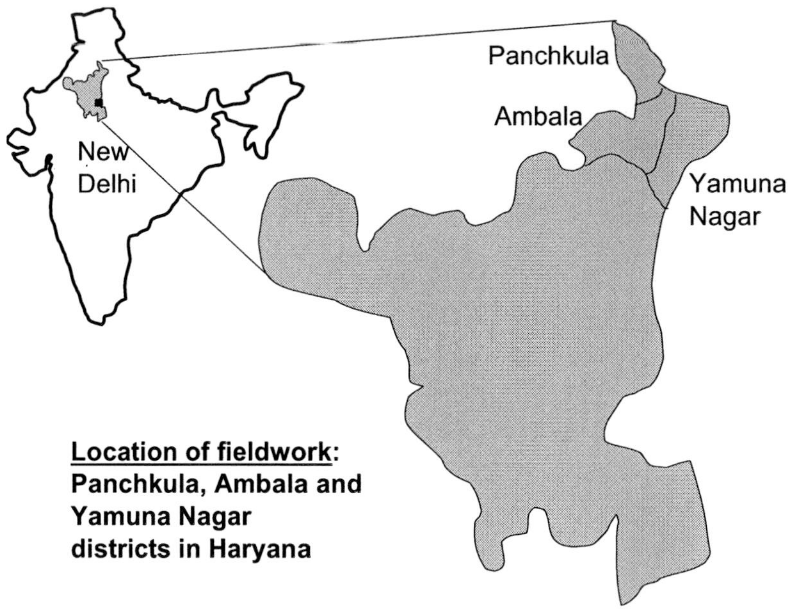

Figure 7.4. Map of districts covered during fieldwork in Haryana.

Dhamala, and Ghidkhala. The *sarpanch* (major of the *panchayat*) resides in Vasdevpura. The impact of the *panchayat*, having a very small budget, on village development is, however, much less than the impact of the Hill Resource Management Society.

In 1996, the people of Sukhomajri were showing signs of prosperity. The survey shows that three out of eighteen families are purchasing foodgrains, while other families can even sell surplus production. Almost all goats have been sold in 1996 and almost everybody owns a couple of buffaloes from which they can sell the surplus milk. Sukhomajri is one of the villages where one dam contains water throughout the year. This makes the village pond appropriate for fish farming, by which means a yearly Rs 2600 is earned by the Hill Resource Management Society.

As shown in Table 7.1, 80% of land (48 ha) is already under irrigation, while the remaining 20% (12 ha) shall be irrigated from 1996 onwards, due to the construction of a government tubewell. The government tubewell is becoming the fifth tubewell in Sukhomajri. The total investment to be borne by the government is about Rs 20 *lakh*s.[11] This tubewell will provide water throughout the day, so that all plots can be irrigated during the peak season of October-November. It is expected that the productivity of land will increase even further, as far as the standing crop is not eaten away by wild boars which have emerged in the regenerated forest.

In Sukhomajri the time is near for harvesting an impressive stock of *khair*[12] trees, which grew because of their effective protection. The improved density of the tree cover has already led to a reduction in the fodder and *bhabbar* yield, which demonstrates the effective-

[11] One *lakh* is equal to hundred thousand.

[12] The botanical name for *khair* is *acacia catechu* (Panda *et al.* [155]).

Figure 7.5. Village gathering on a properly functioning earthen dam in Salehpur in the Yamuna Nagar district in Haryana.

Figure 7.6. Irrigated fields and mango trees (to the right) of Sukhomajri in the Panchkula district in Haryana.

Table 7.1. Sources of irrigation in Sukhomajri.

Irrigation source	Land irrigated (ha)	%
Tubewell number 1	16	26.7
Tubewell number 2	12	20.0
Tubewell number 4	4	6.6
Dam number 2	16	26.7
Total	48	80.0

Note: Tubewell number 3 is used for drinking water only.

Table 7.2. Some socio-economic characteristics of Sukhomajri and Dhamala.

Sukhomajri:	Dhamala:
Small landholding Hindu–*Gujjars*.	Big land holding Sikh–*Jats*.
Fodder grass from forest is very important.	Most fodder is collected from private land.
Caste-wise homogeneous community.	Caste-wise heterogeneous community.
Fairly equal distribution in benefits.	Only landholders are benefiting.

ness of social fencing in Sukhomarji.[13] The people of Sukhomajri have made a claim to share in the tree yield, but this has not yet been granted by the forest department. Distribution of the benefits from harvesting *khair* is potentially a new source of conflict. While Sengupta [183, page 41], argues that in return for their protection they should be given 50 per cent of the sale proceeds, Haryana's notification on joint forest management does not provide such legal guarantees (see Section 2.3.1.). The potential share of the *khair* harvest is expected to be much higher than the yearly *bhabbar* sales.

3.1.2. *Bhabbar* Conflict between Dhamala and Sukhomajri

Dhamala is next to Sukhomajri.[14] Yet, Dhamala differs in many aspects from Sukhomajri. The dichotomous socio-economic differences are presented in Table 7.2.

Since both villages are geographically very close, but socially very different, they have been in conflict with each other repeatedly. After seeing the impact of the dam in Sukhomajri, people from Dhamala also demanded a share of the water from the dam, which was as close to Sukhomajri as it was to Dhamala. To avoid a conflict, a separate Hill Resource Management Society was formed in Dhamala. The forest department constructed two dams in Dhamala. The dams still provide irrigation water, partially through a spillway and partially with the help of a diesel pump.

[13] Studies of The Energy and Resources Institute resource team indicates that the grass yield reaches its maximum after about 6 years of protection and declines somewhat when protection is continued (Panda *at al.* [156]).

[14] More characteristic of both villages can be found in Tables B.9–B.12 on pages 192–194 in the Appendix.

While the old conflict about water could be settled, a new conflict emerged. The Haryana forest department decided that the protected forest should be managed jointly by both villages. While protection of the forest works well, since both villages have shifted to buffalo rearing, leasing the *bhabbar* crop led to a conflict. Both communities have the right to lease the same *bhabbar* crop, while they have to share its yield. The conflict was intensified when Dhamala requested to split the forest.

The influential farmers of the *Jat* caste in Dhamala are solely interested in the monetary value of the matured *bhabbar*. They do not extract grasses from the forest to feed their cattle, as it is sufficiently available from their private lands. They favour a lease of the whole *bhabbar* yield. Playing it the hard way, Dhamala made a deal with the paper factory for Rs 1.5 *lakh*, without informing anybody from Sukhomajri in 1995. Hence, their main purpose is to increase their cash flow into the village fund to develop their village. Earlier, Dhamala only received a yearly nominal amount of about Rs 3500 from Sukhomajri as their legal share. They see this as their first chance to make a good profit with their Hill Resource Management Society.

Additionally, there are many aspects of development in Dhamala, which need investment. For instance, the people of Dhamala want to increase the height of the two dams by 1 meter and they want to improve the spillway. Earlier, the village fund of Dhamala was used for the following:

- Repair of the irrigation dam;
- Repair of the road from the village connecting the main road to the town Pinjore;
- Construction of a boundary wall of the school for safety of the children;
- Putting a door in their *dharamshala* (guest-house for Hindus).

In contrast with Dhamala, the people of Sukhomajri are more interested in free availability of fodder grasses, which sometimes motivates them to cut premature *bhabbar* grasses. In Sukhomajri, where the relatively small landholdings are quite equally distributed, they use fodder from the forest to feed their buffaloes. They are more dependent on the forest than the *Jat*s from Dhamala. For instance, the average level of forest dependence is 0.649 in Sukhomajri as compared to 0.582 for the whole of Dhamala (see Table B.12 at page 194) and 0.515 when we consider *Jat*s only.[15] Splitting the forest means a decrease in availability of fodder for Sukhomajri, which has to be compensated by additional purchases. There are about 40 members who collect fodder from the forest, leading to a collective loss of around Rs 1 *lakh* for Sukhomajri. There is no way to compensate for this loss, even if they would sell more *bhabbar*. Furthermore, the profit of *bhabbar* is put in the village account, while manual collection of fodder has a direct private benefit.

Another reason for the people of Sukhomajri to be against splitting the forest, is that they claim to end up with a smaller share of the forest (40%) albeit of a worse quality. The people of Sukhomajri are only ready to divide the forest, when they obtain 60% of the forest area, so that they are also compensated for the quality difference. However, the forest is geographically closer to Dhamala, while Dhamala is more populated than Sukhomajri. The conflict has already led to a decrease in profits for both communities.

[15]Forest dependence is defined as the amount of goods collected from the forest as compared to the total consumption of these goods. A formal definition is given in Equation A.1 at page 168.

Table 7.3. Present status of dams in Nada.

Dam number	Fishing	Pipeline	Irrigated area (acres)	Note on functioning of the dams:
1	yes	yes	100	Fully functional
2	no	yes	0	Silt: outlet clogged
(3)	no	no	0	Silt: no irrigation
4	no	no	20	Irrigation by engine
5	yes	yes	40	Functions after repair
6	no	yes	15	Silt: 3× repaired
(7)	no	not any more	0	Broken
8	no	not yet	50	Constructed in June 1996
total	25%	50%	225 (acres)	

Note: Dam numbers 3 and 7 are in Harijan Nada.

Source: Haryana forest department [82, page 92], cross-checked with field observations.

Besides splitting the forest, the forest department may decide to close the forest completely for the villagers. This seems unlikely, because the people of both villages have been protecting this forest area for two decades. It would be unfair to deny them any benefits from the forest.

The conflict between Sukhomajri and Dhamala shows that conflicts can also exist between villages. These conflicts complicate successful implementation of the joint forest management programme.

3.1.3. Reduced Benefits in Nada

The experience with the Hill Resource Management Society in Harijan Nada has been disappointing. Excessive rains[16] in 1987 destroyed the *bhabbar* plants. They tried a few replantations, but these were not successful.

Two dams were constructed in Harijan Nada (see dam number 3 and 7 in Table 7.3). Dam number 3 never provided irrigation water, because it silted up soon. It is very unlikely that they can desilt this dam on their own, which costs at least Rs 20000. Their hardship exacerbated after the collapse of their second dam (numbered 7) in 1994. To compensate for the failing irrigation facility, a diesel pump was purchased in 1995, exhausting the village fund. The diesel pump is used by the land-holders without paying water charges to the society, because they already have to purchase the diesel themselves. In 1996, the forest department repaired dam number 7. Unfortunately, it broke once again during the consecutive monsoon rains.

The only chance for them to increase their village fund by themselves is to sit and wait for the trees to mature; provided that they will get a share in the sales of timber. Once they

[16]There was extremely excessive rainfall in 1987: 1450 mm, while the average rainfall is 750 mm per year in that region. Peak rainfall was reported on 16 and 17 September: 415 mm. It caused devastation in the whole of Punjab.

get a share in that harvest as a return for their protection, they can have more financial space in the Hill Resource Management Society and become active again. Otherwise, the only benefits of the forest are through grazing and through collecting fuelwood.

Harijan Nada is very near to Panchkula, the suburban area within the union territory Chandigarh. Many people have a job there. This has reduced the forest dependency substantially. The average level of forest dependence is 0.505 in Harijan Nada (see Table B.12 at page 194), the lowest of the whole sample.

In comparison with Harijan Nada, Main Nada is better off. Out of 6 dams, 4 are operational, irrigating up to 225 acres[17] of private land. The status of Main Nada's dams is shown in Table 7.3. On an average, 14 hours of water is provided to each family. This has enhanced the agricultural output in Main Nada to a great extent.

While the people have become economically more or less independent, the Hill Resource Management Society does not prosper. The present village fund consists of Rs 10,000 which is even too low for maintaining the dams. In November 1996 they will sell *bhabbar* for the first time, but the net profit is not expected to exceed Rs 5,000. The trees in their forest will mature within 2 to 4 years. Earlier they did not share in the timber harvest. They expect to get a share this time. They earn Rs 10,000 per year by leasing dam numbers 1 and 5 for fish farming to a contractor.

The president of Main Nada asserts that the divisional forest officer in Pinjore put a claim on the income from the fish lease in September 1996. In response, the divisional forest officer argued that since they have constructed the dams, they should also have the profit. The villagers disagree, because they argue that they were given these dams in return for protecting the forest. However, since there is no formal joint management agreement between the Haryana forest department and Main Nada, it is unclear who will ultimately obtain the profit of the fish lease.

3.1.4. Bamboo Management in Godam

Godam is on the border with Himachal Pradesh.[18] Kalka, the nearest town, is 8 km away. Godam is about 15 km from Sukhomajri.

> Godam is one of the first villages where a diagnostic study was carried out by a Hill Resource Management Society support team with forest block staff in July 1989. (Gupta [73])

There used to be a persistent conflict between the low caste *Bhanjda* community and the forest department, based on a mutual mistrust between them. This became clear after both parties were interviewed in isolation by people from The Energy and Resources Institute. They also organised several meetings between people from the forest department and people from Godam. Ultimately, the bias between the two parties could be eliminated. Prior negotiation between the forest department and the village community is crucial for the formulation of a proper joint management agreement. The earlier mutual conflict resembles a prisoner's dilemma situation, where both parties played their dominating strategy: defect.

[17] One hectare is equivalent to 2.471 acres.

[18] More details about Godam can be found in Dhar [44], Gupta [73], Kanetkar and Varalakshmi [93], Singh *et al.* [191], and Varalakshmi [208].

The basket-makers extracted bamboo, while the forest department tried to stop them, as they did not see any use in helping the basket-makers. One way of escaping this dilemma is through negotiation. Keeping to the joint management agreement is beneficial for both parties. Mutual participation can be preserved as long as nobody defects.

A Hill Resource Management Society was created on 2 February 1990. The joint management agreement led to a three-fold increase from 300 to 900 pieces of bamboo per year. The yearly fee to the government decreased from Rs 180 to Rs 70. This change led to a more stable source of income for the basket-makers. Earlier, mainly women made the baskets at home, while their men were doing all kinds of temporary jobs outside the village. They also had to travel quite a distance. The increase in the supply of bamboo made it possible for the men to start making baskets as well.

The basket-makers, obviously, have a very high stake in the forest, because their main source of income is coming from the forest: bamboo. It is no surprise that they have the highest level of forest dependence: 0.882 (see Table B.12). Interviews which covers more than 90% of the basket-makers, reveal that the people were effectively protecting the forest by stopping fires and cleaning the bamboo clumps from dry grasses. They are protecting the forest under a traditional system of social fencing. When they find somebody destroying something, they report it immediately to the forest guard.

Besides bamboo, the villagers also collect fuelwood from the forest. The common forest is classified as a *protected forest*. Thus, the people are permitted to extract only their basic needs for private consumption. The extraction of resources is closely monitored by a forest guard who goes from house to house to check the amount of extracted resources. The assistant of the forest guard, who is the treasurer of the Hill Resource Management Society, reported ten cases of illicit entries into the *reserved forest*, twenty cases of people found in possession of too large pieces of wood, leading to fines of Rs 36 to 200, and one case of where one person took five pieces of bamboo too many, who was subsequently fined Rs 60.

During the monsoon, when there is no permit for collecting bamboo, the forest guard provides employment for planting trees in the forest and on road-sides. Trees planted in the rainy season have a better chance to survive than during other (dry) seasons. With the help of this additional activity, the *Bhanjda*s can be employed throughout the year.

Because the main market, in Kalka, is too distant for most of the villagers, and the basket-makers have no capital for storage and transport, they sell their baskets to so-called middlemen. There are four of them. The basket-makers also borrow money from these middlemen. On these loans they do not have to pay a direct interest, but it decreases the prices of the baskets. The reduction in price is equal to an interest rate of 15–40% per year.

Godam is an example of *interlinkage*,[19] because loans are provided only to those people who regularly supply baskets. In that way the middlemen can set the prices of the baskets in their advantage. They have the capital and the storage capacity to wait until the prices are good enough to sell the baskets in bulk. Sengupta suggests that

> the forest department should take initiatives to encourage villagers to market their produce rather than depending on middlemen. (Sengupta [183, page 44])

From the people's side, it can be observed that their income has increased over the last few years. But their present debts are still quite high. Mr Bali Ram, an elderly man from

[19]Basu [12] analyses the interlinkage situation in rural India.

the *Bhanjda* community, was complaining about it:

> Our situation looks fine on the surface, but we are rotten from the inside, because of our debts. (Based on interview with Mr Bali Ram.)

Mr Sita Ram, another member of the Hill Resource Management Society, made the following statement on the impact of the creation of the Hill Resource Management Society:

> Our situation has not improved at all. We are as indebted as before the creation of the society. Loans have to be taken in stress situations, such as marriage, death and illness. The loans have grown to such proportions that we are unable to get good prices for our baskets which we have to sell to the money lenders. (Based on interview with Mr Sita Ram.)

Yet, not all people are indebted. Moreover, most of the people were very enthusiastic about the bamboo availability. Some families are taking up other work as well, like becoming a shop-keeper or industrial worker. By saving money, they managed to repay their debts to the money lenders. They are able to market their products on their own behalf. Some villagers claim that the Hill Resource Management Society will be fully successful only if all members have direct access to the market, without interference of middlemen. This will give them better prices for their baskets, so that all members can repay their debts. The dependence on middlemen is perceived by basket makers as the main burden in the way of their development. This dependence is, however, not easily eliminated because of the vested power with the middlemen, whom would otherwise see their income drop by about Rs 10,000 per year.

3.2. The Ambala District

3.2.1. Water Shortage in Mandpa & Gobindpur

Mandpa and Gobindpur are in the plains, bordering the Morni hills in the Ambala district (the upper-right corner in the Ambala district as presented in Figure 7.4). The nearest town, Raipur Rani, is at a distance of 6 km. There is a large plain between Raipur Rani and the Morni hills consisting of about 400 hectares of private land. This area is shared with two more villages: Ratatibi and Badaur. The land is owned by about 200 families. All these villages are quite homogeneous: most of the people are Hindus, belong to the *Gujjar* caste, and own some land. A single Hill Resource Management Society was registered on July 7, 1989 in Mandpa and Gobindpur (see Table B.1). These two villages are the nearest to the forest, and have traditionally been the main users of the forest.

However, the situation in Gobindpur, where a few bigger landlords reside, is somewhat different from Mandpa. These landlords dominate the Hill Resource Management Society to such an extent that they can secure continued access to irrigation water from the dam. Because of this, difficulties emerged in the equal distribution of water, where the landlords nearly take all the water and the smaller landowners hardly obtain any irrigation water. Land-ownership is quite good in Gobindpur, but water is scarce. Moreover, the region is a drought-prone area. The harvest from rainfed agriculture is very low. The ground water level is about 100 meters deep. As a result, in the dry and hot summer season, due to shortage of fodder and water, the main buffalo, ox and cow population migrates with about one member per family to far off cities like Ropar in Punjab, Haridwar in Uttaranchal or Yamuna Nagar in Haryana. In those cities, their animals can graze near the riverbanks or on the barren land of the landlords residing near these cities. They also find a good market for selling surplus milk, such that they can support their family back in the village as well.

In 1986, a huge dam has been constructed, costing about Rs 10 *lakh*s. This dam has a catchment area of 62 ha and a command of 59 ha, such that it can potentially irrigate 15% of total available agricultural land in the region. The villagers explained that the capacity of the dam has declined, due to heavy siltation, which emerged from the bordering almost completely barren and sandy hill slopes. The people tried their best to solve the siltation problem. With own resources, they cleaned the spill-way and tried to desilt the water storage with the help of a tractor for 3 to 4 years. Their village fund exhausted in the process, while the siltation problem remained. They only treated the effects, without stopping the cause: the barren hill-side in the catchment. Presently about 40 ha of land, belonging to 50 families, can expect water for the winter crop, once in a season. Luckily the dam is still there and it still serves as a source of irrigation water, but the people are disillusioned. They had higher expectations during the construction phase of the dam.

Here, a dam has been constructed first. Three years after the construction of the dam, a Hill Resource Management Society was registered. At that time the storage capacity of the dam had already been reduced substantially due to siltation.

3.2.2. Masoompur

Masoompur village is near to Gobindpur. Both villages are separated by a river. In 1986, two dams were constructed in Masoompur, while the Hill Resource Management Society of Masoompur was registered on 29 May 1987. While one dam, which was constructed in a small stream, broke immediately the next monsoon, the second dam provided water for a few years, but ultimately broke as well. In 1995, the dams were repaired, but they broke immediately during the consecutive monsoon. Nevertheless, irrigation is still possible for farmers in Masoompur, because 120 out of 200 hectares of private land is irrigated, either by a private tubewell or by a channel system.

In spite of the failure of the dams in Masoompur, the society has been quite successful in leasing *bhabbar* grass, which yields about Rs 50,000 per year. Earlier, they invested Rs 50,000 in the construction of a temple. They are planning to build a *dharamshala* in 1997 of which the total cost is estimated at Rs 1 *lakh*.

3.2.3. Thathar, Madhuwala & Dharampurbir

The river which separates Masoompur and Gobindpur leads to Dharampurbir after a half km from Masoompur, to Madhuwala after 2 km, and finally to Thathar after 4 km. Dharampurbir and Madhuwala are two small hamlets with about 12 families. In Thathar reside about 52 families. The villages are remote, as they are neither connected by road, nor by electricity. In 1996, 5 solar energy cells were placed in Thathar, which provides light at night and power to some electrical appliances. Only a few families benefit from this.

After the forest department and the villagers came to a joint management agreement, a Hill Resource Management Society was formed on 7 July 1992 jointly in Thathar, Madhuwala, and Dharmpurbir. These villages could get the lease-right for *bhabbar* as a reward for protecting the forest. The managing committee is equally represented over the three villages: the president, one female and one male managing member are from Thathar; the treasurer, one female and one male managing member are from Madhuwala; the secretary and two male managing members are from Dharmpurbir.

No dams could be constructed in this village, resulting in 62 out of 63 hectares of private lands in Thathar being rainfed only. The society does not charge any fee (*dati*) for the collection of fodder from the forest. People who reside near Thathar traditionally had rights for using the forest in the past. Therefore, the perception of the villagers is that the creation of the Hill Resource Management Society did not enhance their ability to collect resources.

The only income which accrues to the Hill Resource Management Society is by sales of *bhabbar*, leading to a yearly income of Rs 15,000. The present account of the village fund is Rs 16,000. In 1996, Rs 12,000 was spent from the village account on the purchase of tools for harvesting *bhabbar*.

3.3. The Yamuna Nagar District

3.3.1. Bhabbar conflict between Salehpur and the forest department

The concept of joint forest management is based on the interaction of forest departments with local people. However, the interests of people and forest departments can be in conflict, as in Salehpur, a *Gujjar* dominated village in Haryana near the border with Himachal Pradesh. The Hill Resource Management Society of Salehpur, a local organisation constituted by the villagers, has played a key role in joint forest management and pioneered various development initiatives in the village through its novel scheme of leasing out *bhabbar* to generate funds. But, the State forest department's decision to auction *bhabbar* and fodder in the forest land threatens to shatter the villager's dreams.

Salehpur has approximately 130 families, including 106 *Gujjar* and 20 *Harijan* families. The managing council of Hill Resource Management Society, which has nine members, including two women, looks after 377 ha of reserved forest. This land was earlier owned by the villagers but the government acquired it about 20 years ago - to turn it into a reserved forest by paying a nominal fee of Rs 1,050 per ha to the owner. At present, six villages are subsisting on this forest: Mamtapur, Arsulpur, Thaska, Milak, Jhanda, and Salehpur.

Salehpur has two dams. The first was built at a cost of around Rs 10 *lakh* in 1982-83 by the State agriculture department. It started supplying water to 20-24 ha of land from 1983. But the outlet of the dam has been clogged since 1990 by silt from the catchment area. Neither the Hill Resource Management Society nor the agriculture department have taken any steps to remove the silt from the outlet.

In 1991, the forest department decided to build another dam about one km away from the first at an expenditure of around Rs 15 *lakh*. This dam can provide irrigation water for 40 ha of land. In the ensuing excitement over this new dam, everybody forgot the old one.

Desilting the outlet or strengthening the first dam would not have cost more than Rs 1 *lakh*. In 1995, the increased pressure of silt and heavy monsoon rains caused the spillway to collapse. Consequently, the dam burst. The resulting flood washed away fertile soil in four ha of private cultivable land, converting it into a wasteland. The destruction caused by the mishap amounts to at least Rs 10 *lakh*, the total cost of the dam. Besides, reconstruction of the dam would now amount to an additional cost of Rs 10 *lakh*.

Krishan Lal, the *sarpanch* of Salehpur says,

> We do not really need these dams. A good tubewell costs only Rs 2 *lakh* and is much more reliable (than dams) for irrigation purposes. If the government had really cared for the development of the village, it would have dug a number of tubewells. It would have been possible to irrigate the 120 ha of private land in the village with the money invested in the first dam. (Based on interview with Krishan Lal.)

Moreover, the construction cost of the first dam of Salehpur could have been much less. According to S. K. Dhar of Haryana forest department, the third dam of Sukhomajri in Haryana, which irrigates 16-24 ha, was constructed in 1981 - only two years earlier - at a cost of Rs 70,000.

The damburst woke up the people of Salehpur. Fresh elections to the Hill Resource Management Society council were held on October 4, 1995, and the new members undertook everything to avoid a mishap of such a magnitude in future. Krishan Lal,Krishan Lal, who has taken a lot of interest in the Hill Resource Management Society, saw the possibilities of making investments in community development. As the government provides only Rs 10,000 for the development of the village, the Hill Resource Management Society decided to lease the *bhabbar* and collected a stunning Rs 2,65,500 in the village fund, which could be effectively used to make significant improvements in the village.

The Hill Resource Management Society strengthened the second dam with the help of the village fund. As a result, a more equitable distribution of water has been achieved by laying a 1,890 meter pipeline. The water pressure of the dam is twice as high as private tubewells. The dam water can irrigate one ha of land in 10 hours, whereas private tubewells take 20 hours to irrigate the same area. Besides, as the new pipeline has many outlets, many people can withdraw water simultaneously, so that peak demands of water can be met fast and effectively. It also reduces the risk of failure of the wheat crop - the main winter crop.

This is probably why the villagers of Salehpur say that their society has been successful only since 1995. Earlier, the people have sold all their goats. In 1996, even the *Harijan* families are rearing buffaloes.

In order to initiate more development projects, the Hill Resource Management Society needs to lease more *bhabbar*. The main problem in this village is the lack of irrigation facilities. There is an old tubewell in the village which can be used again for irrigating 40 ha of land by deepening the well. This will cost at least Rs 70,000. Land irrigated by a tubewell yields 6.97 tonnes per ha of grains per year. On the other hand, land irrigated by the dam yields 4.08 tonnes of grains per ha per year, while rainfed land yields only 1.22 tonnes of grains per ha per year. The revenue from the yield is Rs 37,483, Rs 21,668, and Rs 8,109 respectively per ha of land irrigated by a tubewell, a dam and rainfall.

The success story, however, has started too late for the forest department. The officials are frustrated as nothing has been reinvested in the forest. They are also doubtful whether all the people are actually getting a share of the benefits from the forest.

In a recent move, the forest department has decided to lease the *bhabbar* grass through an open auction to generate an additional income of Rs 4.51 *lakh*, though the Hill Resource Management Society had already paid Rs 75,500 to secure the *bhabbar* lease. The Hill Resource Management Society would have got at least Rs 5 *lakh* if they had leased the *bhabbar* grass, claims the *sarpanch*. No money has been repaid to the Hill Resource Management Society so far, while the people of Salehpur are denied access to *bhabbar* and fodder grass in the forest.

Normally, the villagers collect fodder from the forest from November to March - after the harvest of *bhabbar* - by paying a nominal fee of Rs 100. But, if the decision to auction the grass is not changed, the villagers will also lose their right to collect fodder. The decision will have an adverse effect on the people of the *Harijan* community. Every family owns at least one buffalo and will have to bear an additional expenditure of about Rs 4,000 for purchasing fodder.

The joint forest management programme seems to be breaking down here. And the people of Salehpur are not getting any reward for protecting the forests. This will only result in their alienation and force them to refuse further participation in joint forest management programmes.

4. Comparison between the Case Studies

The process, which is currently referred to as joint forest management in Haryana, initially emerged spontaneously. Encouraging results in Sukhomajri and later on in Nada motivated the Haryana forest department to extend this model. However, the results of the replication have been disappointing and this period is also called the *dam-and-lease* phase to reflect the lack of institution building. The experiences of the dam-and-lease phase are reflected in the case studies of Mandpa/Gobindpur, Masoompur, and Salehpur. Only when The Energy and Resources Institute entered the scene in 1990 the results improved and the process could be rightly named joint forest management. Moreover, Godam and Thathar are the only cases where a joint management agreement preceded the registration of their Hill Resource Management Society.

An indicator of interest is the additional income from participation in forest management, which was taken as the payoff in games as derived in Chapter 5. In the villages studied in Haryana, villagers depend moderately on protected forests. After the creation of the Hill Resource Management Society, basic needs like fuelwood and one headload of *bhabbar* could be collected freely from the forest. Fodder-grass could be collected from the forest a few months per year as agreed upon by the Hill Resource Management Society, by paying a fodder fee. The people of Godam have access to bamboo as well. In Sukhomajri, Dhamala, Mandpa/Gobindpur, and Salehpur, several villagers get water from earthen dams, constructed in the forests by the government. The highest payoff is found in Mandpa/Gobindpur. This is mainly because of the large amount of water that they can extract from the dam, which adds up to 59 hours per family.

Another indicator of interest is the income of the Hill Resource Management Society. This is the highest in Salehpur, but it is also quite substantial in Sukhomajri, Masoompur and Dhamala (more than Rs 40,000) (see Table B.10).

When the current account of the village fund is taken as an indicator, Salehpur is worst off, because they exhausted their village fund. The inflow into the village fund is limited, as some landowners are reluctant in paying the water charges. Still, others do pay the water charges, but this is generally not sufficient for maintaining the dams. Subsequently, the landless people do not fully benefit from the dams. Furthermore, the income of Hill Resource Management Societies is generally not reinvested into regeneration of the forest. The step to bring the initiative of planting saplings among the villagers could not yet be achieved.

In Godam, Sukhomajri, and Salehpur a full participation in which villagers check each other takes place. Hence, social fencing works well in these villages. These villages also have a high average level of participation (see Table B.12). The forests are still a state property regime in Haryana. This situation is changing. The majority of the people in Sukhomajri, Dhamala and Harijan Nada are claiming that they are either the owners or the joint owners of the forest (see Table B.9). This is in contrast with the other villages where most of the people consider the government as the sole owner. The process of joint forest management in Haryana is still in its initial phase. As the process goes on, probably more people will claim that they are the (co-)owners of the forest. Hence, a transition from a state to a common property regime is going on.

In Haryana the forest department has taken a real interest in regenerating the forest. This example could be followed by other States in India. It is important to come to a joint management agreement first to avoid the type of conflicts about resource sharing which occurred in Sukhomajri, Dhamala and Salehpur. It is also important in reducing the siltation problem in the catchment area of the dams.

Wealth and participation can be connected with forest dependence and education. People who depend a lot on common property regimes generally own a little: they are poor. They participate more in forest management, because they have high stakes in its preservation. This is an essential link between common and private property regimes (see Figure 7.7).

Figure 7.7 gives a schematic overview of the linkages between common and private property regimes. It shows the interrelation between two common resources: forests and dams, and three private resources: cattle, land, and labour. These common and private resources are linked by their benefits streams. The resources become property when one has the legal right to claim these benefit streams (see section 3.1.).

5. Conclusions and Recommendations

In this Chapter, I have compared the process of involving people in forest management in Haryana with the actual performance at the village level. The main restrictions in implementing joint forest management in Haryana are the conflicts which persist between villages, and between a village and the State.

Which lessons can be learnt from the current experience of joint forest management in Haryana? The main reason for success in Sukhomajri is that the Chandigarh administration, through the Central Soil and Water Conservation Research and Training Institute, had high stakes in changing the behaviour of the people in order to reduce the siltation level of the Sukhna lake. The only way to win the confidence of the people concerned was to build a

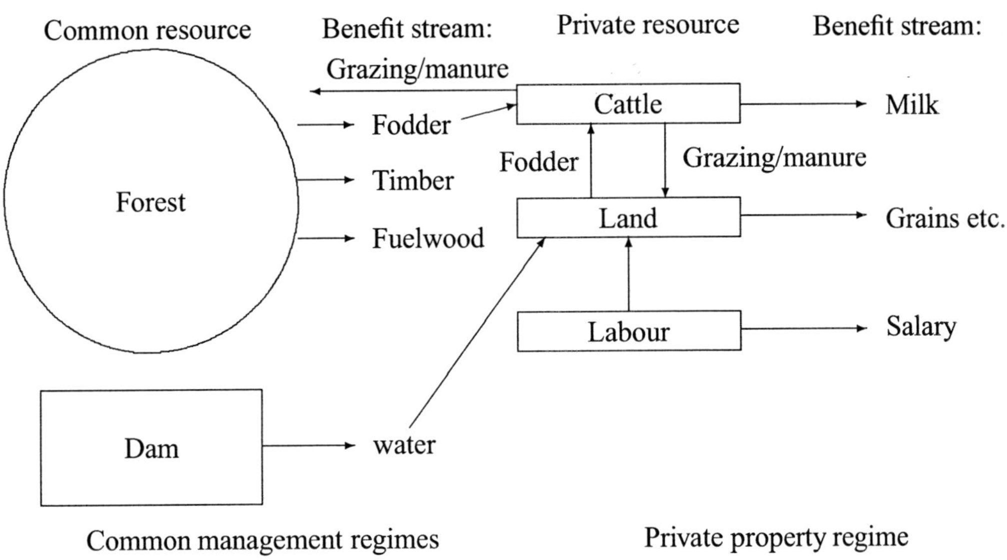

Figure 7.7. Linkages between common and private property regimes.

demonstration dam. In other villages, there is no Sukhna lake which is silted up through free grazing practices. Unfortunately, the same attention and will-power have not been provided by the State in other villages in order to replicate the model in Sukhomajri. This may be an important reason why replication of the Sukhomajri model has not become very successful.

A few guidelines for a better chance on replicating the Sukhomajri story are listed below.

- Equity of distribution: everybody should get a share in the benefits. It has been very difficult to establish an equal distribution of water, because of unequal sizes of landholdings and unreliable supply of water. Water rights are provided to each villager and water charges are ploughed back into the village fund. However, the additional income due to irrigation goes to the landowner. A profit-sharing mechanism would have been more equitable, but is possibly more difficult to negotiate.
- Interaction between the villagers and the forest department should lead to a joint management agreement, which is a match of interests between both parties. Preferably this joint management agreement should come about before the construction of irrigation dams is undertaken. Otherwise the villagers are unlikely to take responsibility in maintaining this State-built dam.
- Meetings of the executive committee of Hill Resource Management Societies should be encouraged, to strengthen the local institution. Maybe some meetings should be attended by government representatives in order to have an optimal participatory environment.
- It is necessary that the catchment area is quite 'green' before a dam is constructed, to reduce the chance on siltation and dam breaches. Hence, in addition to the joint

management agreement it is also important to create an environmental optimal precondition.

Closing the forest by converting (free) grazing into stall-feeding (selling goats and purchasing buffaloes) does not always lead to a desirable solution. This argument goes again the outcome of a study by the Varalakshmi [209], where buffalo rearing came out as the economically preferred alternative. Their study, however, ignores that buffalo rearing generally needs a much higher capital and labour input. Alternatively, goats are a source of security for the poor. The veterinary, purchase and labour costs of goats are much less. It is not necessarily true that buffalo rearing is the best choice for the risk-averse poor peasant. Grazing in the forest is not always bad, but *over-grazing* is bad. There are even cases in Madhya Pradesh where the forest growth declined due to lack of grazing. Only if the economic status of the poor peasant improves, they might shift to buffaloes. Only when the people are able to make motivated choices, the real spirit of joint forest management can emerge.

It is also not true that people are always in favour of constructing dams as in Salehpur. There, a tubewell would have been a cheaper and better solution. This consideration could also be included into the policy of joint forest management.

The time horizon within which the benefits of participation accrue to the people is very important. As long as people are continuously benefitting, as in Haryana through access to fuelwood, fodder, and dam-water, participation is likely to be sustained. These benefits tend to decline over time. Protection generally improves the density of the forest cover, reducing the availability of fodder. The actual number of dams which can be effectively used for irrigation on the long run also declines. As it is shown in Tables B.1–B.2, in 1996, 20 out of 36 dams provided water for irrigation. Many dams broke, silted up, neither had a spillway, nor a water distribution system. This limited the availability of water for irrigation. In order to guarantee a continuous benefit stream, it could be considered to let Hill Resource Management Societies share the timber crop in the long run. To facilitate this, there is an urgent need to establish legal rights for people to share forest products. This may also reduce the number of conflicts between the people and the Haryana forest department.

Replication of the Sukhomajri-model has been the main challenge for many years. The initial attempts of the forest department were disappointing, as they were too much focussed on building dams. A more people-oriented approach to enhance institution building was introduced with the help of The Energy and Resources Institute. This has led to better results. However, in some instances the forest department has started to play the defect strategy again: they want the fish lease in Nada, and put the *bhabbar* in an open lease in Salehpur, while the people of Salehpur already purchased the lease right. Will the people retaliate? Will they react in a tit-for-tat manner? The disfunctional forest department is a major threat to the successful continuation of joint forest management in Haryana. In order to stay in the win-win situation, mutual trust needs to be sustained both from State and the villagers.

The game between the State and the people has the flavour of a repeated prisoner's dilemma. There is always a temptation to defect, as this leads to a higher benefit in the short run. However, it threathens the existing and fragile institutions of people's participation, which may break down in the long run. It appears that the people have a longer time horizon than the State.

Chapter 8

Afforestation on Pooled Private Lands: *Chakriya Vikas Pranali* in Jharkhand

1. Introduction

This Chapter presents a qualitative study about the performance of the *chakriya vikas prenali* initiative in Jharkhand. This initiative has led to a high level of participation of the local people, due to a prenegotiated sharing mechanism, where 30% of the harvest yield is saved for the development of the village. The yield is generated on pooled private land and labour from the same village by an intensive multitiered cropping pattern, in a transition towards sustainable private forests. Funding limitations, social unrest and marketing blockages by the State are in the way of rewarding this high level of people's participation in rural India.

In many places in rural India, degraded wastelands can be found. Private wastelands remain uncultivated, when the owners either have no financial resources, or are not motivated to undertake the necessary investments to restore the biomass to such an extent that the land can become productive again. Large farmers (i.e. landlords) focus their effort mainly on fertile land and they are not interested in investing in restoration of marginal or degraded lands. However, these otherwise unused private wastelands can be pooled among various landowners to create a common pool of land. On this pool a multitiered cropping pattern can be initiated with the aid of outside assistance to make the formerly degraded land productive again, such that it may benefit all the poolers in the longer run. This common pool of private land can be managed as a 'group' private property regime, where land-ownership stays with the individual, while its destination for cultivation is based on decision making at the community level.

One such community oriented programme where people have started to pool private (waste)lands at the village level can be found in the State of Jharkhand. Jharkhand has been formed officially on 15th of November 2000, with Ranchi as the State capital.[1] Earlier, Jharkkand formed the Southern region of Bihar.

Encouraged by his early experiments in Sukhomajri with people's involvement in forest

[1] See also: http://www.webindia123.com/states/index.htm

management, Mr P. R. Mishra continued the experiment in the more challenging setting of Jharkhand. The purpose was to replicate the principle of social fencing as previously developed in Sukhomajri (see Chapter 7). This turned out to become a major challenge, since the field realities in both States are quite different. For instance, the State took the responsibility to replicate the Sukhomajri model in Haryana. In Jharkhand, the State was left out willingly and completely. The experiment of replicating the system of pooled land management in Jharkhand is known as '*chakriya vikas pranali*' (cyclic development system) or '*chakriya vikas yojna*' (cyclic development planning). In the *chakriya vikas prenali* model, the objective is to enhance the development of a poor village, by motivating the people to pool private resources, like land, water, and labour, for starting a multitiered cropping pattern on the collective resource pool at the village level. As this process goes on, the marketable surplus is expected to increase year by year. Equitable sharing of the benefits among all participants should strengthen the newly created institution at the village level and boost social capital.

A brief layout of the Chapter is given here. The main principles of the *chakriya vikas prenali*, which seeks people's empowerment at the village level, are discussed in Section 2., to describe its context and basic features. I present village-wise studies based on interviews in villages under *chakriya vikas prenali* (Section 3.), to find the attitude of the people towards the process in their village. The village level experiences are compared with each other in Section 4., to identify the restrictions that are in the way of implementing the idea of the *chakriya vikas prenali*. Section 5. gives general conclusions.

2. Main Principles of *chakriya vikas prenali*

In this Section the main principles of *chakriya vikas prenali* are discussed, to describe how it ought to function.

2.1. Historical Background and Institutional Context

The experiment with forest regeneration through people's participation in Sukhomajri (see Chapter 7) has been replicated in Haryana State and elsewhere. Known as the architect of social fencing, Mr P. R. Mishra, took a chance to continue the experiment in the more challenging setting of Jharkhand.

In 1984, the experiment with *chakriya vikas prenali* started in the villages Khapia and Chhechani (SHRMS [188] and Sinha [195]). In the period from 1984 to 1986, about 2 *lakh* plants were planted on commonly pooled land in Khapia and Chhechani, with active people's involvement. However, at the time of harvest, the forest conservation law did not allow the people to share the harvest of the planted trees. This conflict led to a heated argument between the Bihar forest department and Mr P. R. Mishra. In August 1990, the government of Bihar ultimately decided that the people could fully share the harvest of the trees, which are planted by themselves on their private lands. The government of Bihar officially approved the so-called '1:1:1' sharing mechanism, where one third goes to the land-pooler, one third goes to the protector, and the final third part goes into the saving account of the village, which is known as the village fund. Hence, a change in formal rules was needed to support the pooling process.

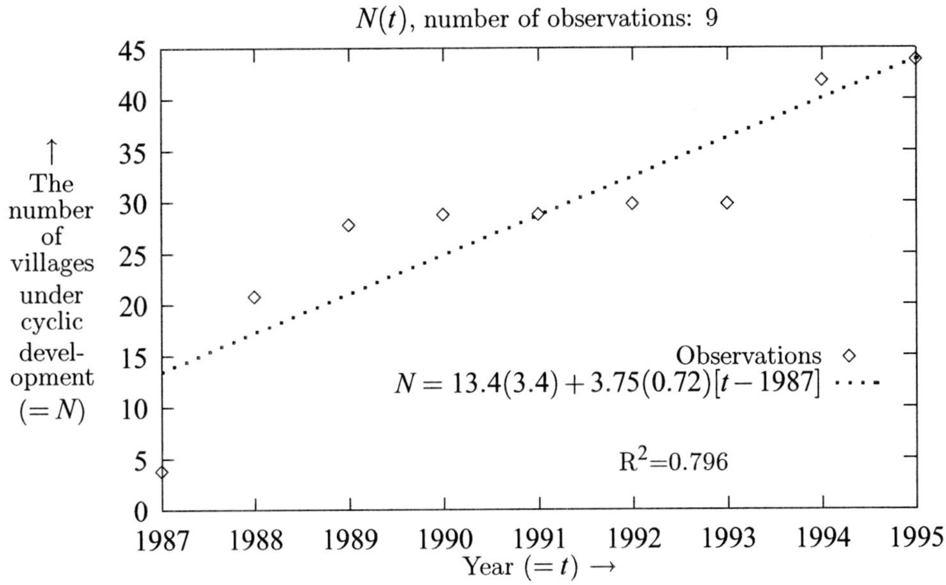

Figure 8.1. Number of villages under *chakriya vikas prenali* over the years.

Note: Standard errors (SE) are denoted in the brackets.

Unfortunately, the acceptance for sharing the trees among the villagers came too late for Khapia and Chhechani. The uncertainty about whether they could share the harvest in the village or not, led to a breakdown of participation in these villages. The plantation was cut by the villagers, to make sure that the Bihar forest department would not get any share. This illustrates that it is important and difficult to back up institutional change at the village level with institutional changes at the State level.

Between 1984 to 1986 a regular interaction between the villagers of both Sukhomajri and Khapia was initiated as well. A group of four Indian forest service officers assisted Mr. P. R. Mishra during the initial efforts. To avoid an institutional breakdown of people's participation in future, the initiators of *chakriya vikas prenali* started working as a registered non-governmental organisation in 1986, known as the Society of Hill Resource Management School. The experiment was continued with a grant from the Society for Promotion of Wastelands Development in five new villages: Bhusaria, Chapri, Karmahi, Khura and Lanka. The focus in these villages was on planting fruit trees.

With additional funds coming in 1988, mainly from the National Wasteland Development Board (66% of total funding), the Ford Foundation (18%), and the Society for Promotion of Wastelands Development (12%) (Sinha [194, page 16]), *chakriya vikas prenali* was extended to twenty-five other villages. Over a period of six to eight years, about 46 *lakh*s of plants were planted on commonly pooled land in thirty villages, while a total fund at the village level of Rs 10 *lakh*s was generated by 1994 (Sinha [194]).

Figure 8.1 shows how the number of villages, where planting started under the *chakriya vikas prenali*, has increased over time. The number of villages increased fast between 1987 and 1989, when the first thirty villages were covered. In 1994, a second increase is observed when the *chakriya vikas prenali* is initiated in another fourteen villages.[2]

In 1996, twenty-five 'mature' and thirteen 'infant' villages were subject to *chakriya vikas prenali* in Jharkhand, while five 'mature' villages and one 'infant' village were subject to *chakriya vikas prenali* in the Aurangabad district in Bihar. During the initial period in the new villages, the poolers should learn extensively from the experiments in the mature and infant villages. The residents from the mature villages are teachers for the new villages.

The Palamau district of Jharkhand, where most of the villages under the *chakriya vikas prenali* are situated, is part of the Chhotanagpur region and used to be dominated by forests.[3] The forests of Palamau are traditionally inhabited by tribals, also known as scheduled tribes. Forests have been depleted excessively over the last few decades and the total area of dense forests is decreasing at an alarming rate. Consider the following quotation:

> As recently as the '60s, superbly well stocked forests covered about *92 percent* of the district's forest area. It plummeted to 41 percent in the '90s. Now the annual forest depletion rate averages about 5-7 percent. (Kadekodi [86, page 29])

The causes for rapid depletion are many. Increased demand in combination with hardly any State control, led the people to trade wood, whether it was allowed or not. Jharkhand is also categorised as the poorest State of India. More specifically, Palamau again is one among the poorest districts of Jharkhand.

It will be a formidable task to realise a change in behaviour of tribals. Tribals are traditionally dependent on the forest and were used to make a living by using and selling forest products like fuelwood, timber, bamboo, fruits, flowers, leaves, honey, vegetables (wild roots found somewhere deep in the forest are used for that purpose), grasses, medicinal herbs, tree barks, etc. About 30% of the forest produce is used for private consumption, the other 70% is leaving the State (Kumar [102]). Through collection of all these products, tribals were self-reliant for centuries. Presently, tribals are able to sell only a small amount of wood or other forest products in the hardly and badly functioning neighbouring markets. By that means, tribals are also able to buy other non-forest products for their own consumption. Due to large-scale deforestation in the region, many tribals either saw a reduction in the availability of forest products or lost access to the forest completely. As a reaction to this process, the tribals either look for labour opportunities, although they are highly uncompetitive,[4] or they seek a new opportunity to apply their natural knowledge of living in harmony with the forest. *Chakriya vikas prenali* is trying to revitalise their traditionally strong experience of the latter case.

---

[2]See Table B.4 at page 187 in the Appendix for the names of the villages, the period of planting, the number and kind of trees planted, the number of water-harvesting structures constructed, and the local fund in 44 villages under the *chakriya vikas prenali*.

[3]For an in depth study on Palamau refer to Kadekodi [87]. Kadekodi *et al.* [86, 88, 89] have a special focus on cyclic development. The general process is also treated by Mishra [137, 140] and Tiwari [203], among others.

[4]Still, the law is on the side of the tribals by reserving some governmental jobs for them.

2.2. Basic Features of *chakriya vikas prenali*

2.2.1. Initiation

The head-office of the Society of Hill Resource Management School (SHRMS) is in Daltenganj and there is one smaller assisting office in Barwadih which is about 30 km away from Daltenganj (see also Figure 8.3). Each office is headed by a project coordinator. Each project coordinator is assisted by four programme officers. The programme officer has to motivate, guide and convince new villagers about the benefits of entering *chakriya vikas prenali*. As in Haryana, the outcome of the negotiation process between the programme officers and the villagers is important for a solid start of the process. This task was performed effectively by the patron, Mr P. R. Mishra, during the initial experiments with the *chakriya vikas prenali* in 1984. The interaction with the villagers plays a crucial role in motivating the villagers to pool land.

The pooling process works as follows. When some villagers are ready to pool a part of their private land, a saplings nursery is initiated in the pool, while the Society of Hill Resource Management School is taking responsibility for providing funds to purchase necessary inputs, like seeds, manure, and labour. Effectively, money for initiating a plantation is withdrawn from an account at the village level which is jointly managed by the people from the Society of Hill Resource Management School and the villagers. This system is set up to stimulate mutual participation and trust among the persuaders from the Society of Hill Resource Management School and the resource owners of the village.

Early experiments with *chakriya vikas prenali* show that it is easier to motivate a villager to pool land of an inferior quality, since they cannot lose much and they enter only a low risk. Nevertheless, a number of such contributions can establish one large, generally jointly clustered, pool of land. This pool of land can be cultivated, while profiting from economies of scale. When such a pool can be created, new employment opportunities appear as well for under- and unemployed villagers. Some people can be employed full-time in the nursery to prepare seedlings and saplings to be planted the following year. They work for about 300 salariable days per year. These full-time labourers are also known as *stipendiary students*. However, the number of people who can be employed full-time in the nursery is limited, which is generally only about 10% of the villagers. A greater number of people needs to be employed part-time during the plantation period, which needs to be done within a relatively short period of time and it is quite labour intensive. Still this type of employment does generally not stretch beyond 100 salariable days per year. In addition, all villagers are referred to as *students*. The pool should be protected by everybody, following the principles of social fencing, which originated in Sukhomajri.

2.2.2. Construction of Water Harvesting Structures

In addition to social fencing, an increase in the productivity of land of inferior quality, also referred to as wasteland, is sought by constructing water harvesting structures, using traditional technologies, well known to the tribals: tie-ridging, ring-ridging, strip trenches, check dams, small ponds, bonding *nallas* and deep wells (Kadekodi [86, page 29]). Mr P. R. Mishra put this quite firmly in words as follows:

> The total amount of rain should be arrested as it falls, where it falls. Not a single

drop should be permitted to leave the village border. (Based on interview with P. R. Mishra.)

Water harvesting structures increase the absorption of water by land and reduce water runoff. In some villages, nurseries are irrigated by a pump-set, purchased with assistance of the Society of Hill Resource Management School, to commence *chakriya vikas prenali*. Water is a prerequisite for making the nursery work successful. Planting is done during the rainy monsoon season so that the new saplings can absorb natural rain-water adding to the water harvesting strategy; provided, of course, that a heavy downpour does not destroy the newly planted saplings and constructed water-harvesting structures.

2.2.3. A Multitiered Cropping Pattern through Agro-forestry

An important feature of *chakriya vikas prenali* is that a mix of many plants with different gestation periods is planted side-by-side in the pool. This is a typical characteristic of so-called agro-forestry. In this manner a forest can be created, while the people can still grow their basic needs, like grains and vegetables.

While it is possible to distinguish three tiers, namely having short, medium and long gestation periods, the actual plantation in the pool generally consists of a near continuum of maturing plants. Some plants even take a life-time to mature, like teak.

Tier 1 consists of plants which have a gestation period of less than one year. Examples are the common agricultural crops:
- **grains,** like maize, rice, and wheat;
- **vegetables,** like potatoes, peanuts, carrots, onions, pumpkins, tomatoes, cucumbers, French beans; and
- **fruits,** like melons.

Tier 2 consists of plants which have a medium term gestation period of more than one and less than about seven years. Examples are:
- **(short duration) fruit trees,** like papaya, banana, guava, and citrus (orange, tangerine, lemon);
- **bamboo,** and
- **(short duration) cash trees,** like eucalyptus, khair, and acacia.

Tier 3 consists of plants which take more than seven years to mature. Examples are:
- **(long duration) fruit trees,** like mango, pomegranate, mulberry, *aunla* (myrobalan), jackfruit, jujube plum, and *mahua*;
- **fuelwood trees,** like gamhaar, subabul, and sisam;
- **(long duration) cash trees,** like teak, cashew, and margosa & neem tree.

Some fruit trees have very long gestation period like *mahua*, of which fruits are being used for wine-making and the residuals of the wine-making process can be used for baking *rotis* (Indian bread).

To commence the pooling process, typically a mix of saplings and seedlings from the three tiers is planted in an area of about 20 acres in the pool of the village. In the first year, the villagers only obtain benefits from the first tier. However, as the pooling process continues, benefits from the second tier also emerge. After a while, the benefits from the first tier gradually vanishes, because of the increase in size of the plants of the second and the third tier, which leave no space for growing plants from the first tier. The needs for products from the first tier can then be met by expanding the plantation, while the benefits from the second and third tier are much higher than the benefits from the first tier. The main idea behind the three-tier scheme is to ultimately meet all possible needs of people needs, namely food, income, housing needs, and so on, sustainably.

2.2.4. Sharing of Benefits

The sharing system is a very important feature of the *chakriya vikas prenali* for persuading the people to pool their land and to keep the process going. *Chakriya vikas prenali* considers the village as the basic unit of development. The sharing mechanism of the produced and marketable output is pre-negotiated and is expressed in percentages at the village level.

The exact amount of shareable output depends crucially on the following factors:

- The strength of the local institution.
- The extent to which the people protect the plantation.
- The extent of consumption of products from the pool by the villagers.
- The faith of the people in being guaranteed a stream of benefits from the pool of land.
- The social structure of the people in the village.
- The extent to which they can keep away from politics in implementing *chakriya vikas prenali*.
- The quality of the land.
- The geographical position of the pool.
- The number and kind of trees which have been planted.
- The development of the trees in the plantation with respect to density, the girth and height of the tree.
- The way the plantation in the pool has been organised.
- The relation with people of surrounding villages.

The impact of these factors differ from village to village and depends on the effectiveness of the village meetings and the extent of participation of landlords and (stipendiary) students. To address these differences, village-wise studies are required, which is done in Section 3..

Initially, a '1:1:1' sharing mechanism, based on a principle of equality, was employed in all the villages. The marketed surplus, once converted into money, should be divided equally over three funds.

1. A land-pooler fund;
2. A student fund;
3. A village fund.

More recently, as a general principle, it was decided that 10% of the marketed surplus should be put aside in a so-called welfare fund. The idea was to return this to the funding agencies. It was expected that the complete fund could be repaid after about ten years. The funding agencies however, not used to having their funds returned, suggested to maintain the welfare fund and to use it in future for replicating the process in other villages. The remaining 90% is shared as before. The latter sharing mechanism is also referred to as the '3:3:3:1' sharing mechanism.

In general, the land-poolers obtain their share instantly, according to simple sharing rules. For example, guava trees can be grown on land of several land-poolers, but the final marketable harvest is pooled. It is not always possible to know the exact number of guavas for each land-pooler. To simplify sharing, the size of land can be taken as a unit, or as agreed upon among the land-poolers.

The sharing mechanism accounts for a village level transfer of money to four parties. Assigning the exact share to each individual is one of the important tasks of the *gram samaj* (village society). Conflicts about the proper sharing of the benefits should be resolved during the general body meetings of the *gram samaj*. The setup of the *gram samaj* is discussed in the next Section.

2.2.5. Evolution of the *Gram Samaj*

Two main institutional changes take place while initiating *chakriya vikas prenali* at the village level.

1. The decision on the use pattern of the common pool of private land is transferred from the individual to the community.
2. A village level organisation is created, known as the *gram samaj*, which consists of a number of committees and subcommittees, each with a specific task, to manage the common pool of land.

The *gram samaj* is the organisational form at the village level to implement *chakriya vikas prenali*. The general body meetings of the *gram samaj* are meant for all villagers and held only a few times per year.

The executive committee, which meets as frequent as every fortnight, forms the main body. It has elected posts of the president, the secretary and the treasurer. During these meetings the stipendiary students receive their stipend. In 1996, the students received a daily stipend of Rs 25. Out of this, 10% is put aside in the *student fund*. In this manner, the stipendiary students are taught to save money. After marketing the output, the 30% share for all the students is also added to the account of the student fund at the village level. The student fund is under the control of the *operator student fund*. The members of the executive committee cannot withdraw money from the student fund without agreement of the operators of the student fund. In a number of villages the student fund is mainly used for the provision of loans.

The village fund is under the control of the *operator village fund*. The operator of the village fund looks after the *village fund*, which accumulates as the 30% share in the marketable surplus. The Society of Hill Resource Management School immediately transfers the grant which is meant for the village to the village fund, on the basis of attaining certain objectives, such as an x number of trees to be planted and a y number of water harvesting structures to be constructed. In this manner the *gram samaj* can decide when to withdraw which amount of money to achieve these targets. Hence, after the money transfer, the responsibility is totally in the hands of the executive committee.

Once the village fund is large enough, it can be used for development projects in the village, like providing drinking water, irrigation facilities, schools, and roads, but also for

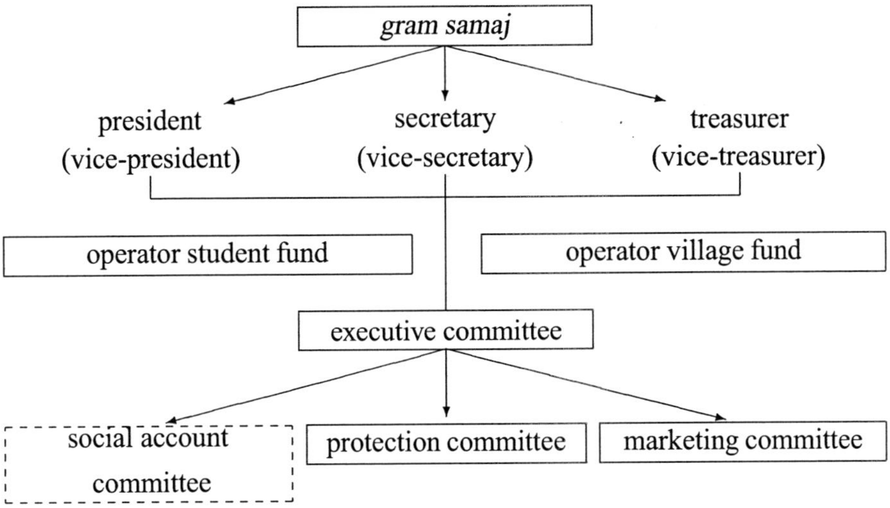

Figure 8.2. The organisational setup at the village level: *gram samaj*.

maintaining the pool of land to maintain continuity in the stream of benefits for the participants in the long run. The village fund can also be used for generating various kinds of industrial activities at the village level, like rope-making from elephant grass, fuelwood-making from subabul, soap-making from papaya, brick-making from clay in nearby rivers, and so on. As before, the members of the executive committee cannot use the money from the village fund without agreement of the operator of the village fund.

Besides the executive committee, there are a number of elected subcommittees.

- The social account committee;
- The protection committee; and
- The marketing committee.

The *social account committee* looks after proper management of the open account books in the village. This committee only exists in about 50% of the villages. The *protection committee* has the responsibility to ensure that social fencing works. They are the guards and are supposed to report illicit extraction from the pool, to decide on further action during the meetings. The *marketing committee* only becomes active after a few years of functioning, when the society can produce a marketable surplus. The link between all such committees is shown in Figure 8.2.

The meetings have become a means by which the villagers can communicate with each other. The meetings of the executive committee enable the people to participate in monitoring and evaluating the plantation activities. During the meetings, villagers can learn about *chakriya vikas prenali*. The meetings are meant for settling intra-family disputes and to reduce discrimination based on caste, creed and gender, in order to improve the social relations in the village. More technical issues, like encouraging the growth of saplings in the pool, are also discussed during the meetings. In short, by pooling resources, the villagers are investing in the future, where the pool of land is the imaginary account of the villagers. The first installments are the saplings, while they hope for a high return in the future when they can sell the matured trees of the third tier.

The following rules and responsibilities characterise the functioning of the *gram samaj* with respect to *chakriya vikas prenali*:

- Every villager should protect the pool by contributing to social fencing.
- Every animal owner should stop the grazing of his cattle in the pool and, preferably, should start stall feeding.
- Every landowner can pool a part of his/her private land.
- A selected number of villagers can do the work in the nursery. Other villagers can help in planting saplings, constructing tanks, tie-ridges and other water-harvesting structures, and all other work related to *chakriya vikas prenali*.
- The poolers should market the surplus with the best of their efforts and the yield should be shared among the participating parties in the agreed-upon sharing mechanism (3:3:3:1).
- Everybody has the right to become a candidate for the elected posts of president, secretary and treasurer, if they can do the work.
- Every adult, irrespective of caste, creed or gender, has the right to vote.
- The executive committee should meet fortnightly to distribute the stipends and to evaluate and discuss the progress of *chakriya vikas prenali*.
- Village meetings of the *gram samaj* can be organised to settle local disputes and conflicts to improve the social relations in the village.
- Outside assistance from the programme officer and the project coordinator to settle disputes and conflicts can be expected, if necessary, but the primary objective is that the people solve their conflicts by themselves.
- The executive committee should decide on the activities to be undertaken by the students and how to spend the available funds, with the consent of the operator village and student fund.
- An open account book of the village, student, welfare and land-poolers fund should be maintained.

3. Village-Wise Case Studies

Implementation of *chakriya vikas prenali* is discussed for villages in the two main subdivisions in the Palamau district: Barwadih and Daltenganj. Different field realities are pointed out in six different villages. The focus is on the experience of the older villages in their struggle to continue *chakriya vikas prenali* without further funding. Figure 8.3 presents a map of the location of the fieldwork undertaken in Jharkhand. Figures 8.4, 8.5 and 8.6 provide three illustrations of the situation in the field.

3.1. Barwadih Subdivision

3.1.1. Chapri and Karmahi: Participation is Strengthening

Chapri and Karmahi are two adjacent villages at a distance of about 5 km in Southern direction from Barwadih. As mentioned earlier, a nursery was set up as early as in 1986. Planting was done until December 1994 in three plots with a total size of about 200 acres.

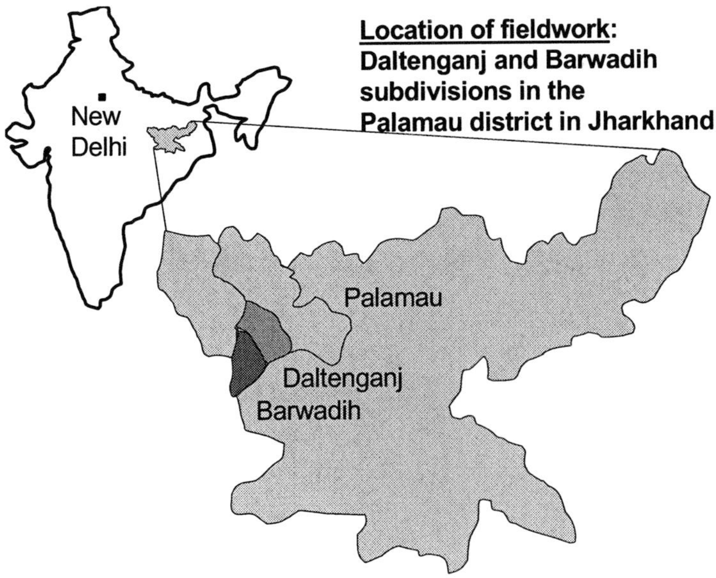

Figure 8.3. Map of subdivisions covered during fieldwork in the Palamau district in Jharkhand.

However, it only succeeded in one plot. The presently still existing pool is owned by 13 persons with a total size of 40 acres, while 20 acres, one tenth of the initial plantation, is still covered with trees (see, for instance, Figure 8.5).

In spite of the somewhat disappointing situation in 1996, the people of Chapri have accumulated the highest local fund of all villages under *chakriya vikas prenali* (consider, for instance, Tables B.3 and B.4). In 1994, they had Rs 116,238 in the student fund and Rs 79,471 in the village fund [194]. While the student fund is used for providing loans, it has been decided to purchase a diesel pump with a pipeline of about 200 metres for Rs 32,000 from the village fund in 1997 in order to irrigate about 60 acres of land. The high amount in the local fund was possible by a decision in a general body meeting of the *gram samaj* to put all the shares of the students in the student fund. The student's share from selling matured trees are expected to be distributed among all households.

For instance, in November 1996, 58 trees (subabul) of the pool were sold. This yielded Rs 11,000. After subtracting Rs 1,000 as cutting cost, the remaining Rs 10,000 was shared according to the 1:3:3:3 sharing mechanism. Rs 1,000 was put in the village welfare fund. Rs 3,000 was shared among 8 landowners who owned the land from where the trees were cut. The shares were given according to the number of trees. Table 8.1 shows this.

Another Rs 3,000 was shared equally among 36 students irrespective of the amount work done for the pool. They opened an account for each student and deposited Rs 83.33 to ascertain that the money would not be spent immediately. In future they will transfer the shares straight to their accounts. The remaining Rs 3,000 was put in the village fund. It will be used for repairing one of their check-dams.

The amount earned may not be very impressive, but it is remarkable that all took place without any outside assistance. Only after the sharing had taken place, the villagers in-

Figure 8.4. Village level meeting in Chapri in the Palamau district in Jharkhand.

Figure 8.5. Fieldguard in Chapri in the Palamau district in Jharkhand.

Figure 8.6. Well-functioning pond to arrest water in Chapri in the Palamau district in Jharkhand.

Table 8.1. Sharing of the sales of trees among land-poolers of land.

Name of landowner:	Number of trees	Share from pooling (Rs)
Samuel Sanga and Lucas Soy	8	414
James Peter Horro	3	155
Paulus Topno	2	103
Biswanath Ram	8	414
Kamesh Ram	8	414
Ram Ratti	12	621
Barosa Ram	9	465
Jeetu Ram	8	414
Total	58	3,000

formed the project coordinator of Barwadih. It was, however, very difficult to obtain permission for cutting these trees. The whole procedure of going up and down in the bureaucratic system of the Bihar forest department took about a year. This demonstrates the bureaucratic constraints in selling the generated output. In the nearby future, it will probably become even more difficult once a large number of trees is fit for marketing.

The secondly created nursery in Chapri has become an impressive oasis in the village. They have created a wealth of trees, tanks and wells (for an example see Figure 8.6). None of the landowners even think about withdrawing his part of land from the pool. One of the main reasons why this part of the pool survived was due to the construction of 5 tanks and 18 check-dams, so that the trees even could survive the most extreme drought situations. This is contrary to the experience with the first nursery, where social fencing broke down, due to a drought. Sukat Ram, the owner of the land in the first nursery, is still ready to pool his land, but as the tree-cover has gone he has started to cultivate his land again with standard crops. He also mentions that the meetings used to be weekly during the initiation of *chakriya vikas prenali* in Chapri with more than 150 people present. The number of meetings has reduced considerably.

There are still a number of people in Chapri who are ready to pool their land, but they did not get the chance. Uday Paswan is ready to pool 10 acres, because he finds the benefits from pooling very attractive.

As the pooling process continued in Chapri, it turned out that 0.5 acres of land belonging to a landlord living in Barwadih had been included into the pool. This landlord has been interested in attending the meetings, but he has not been invited. He belongs to the *Brahmin* caste, while all the other villagers are tribals. The villagers are afraid that when *Brahmins* start to interfere, they walk away with the benefits. It shows the difficulty in maintaining the principle of equality in caste, creed and gender.

People of Chapri are also eager to produce country-made bricks using the freely available clay from a river which is next to Chapri. By this means they can create about 130 labour days per person per year. There is already one such factory in Barwadih. Only Rs 25,000 is needed to initiate this small-scale brick-making factory. They presume to be able to produce 2 *lakh* stones which will sell for Rs 300-400 per 1,000 bricks. A committee of five people is ready to take the responsibility. The wage-rate should be put at Rs 50 per 1,000 bricks. One person can make about 1,500 bricks per day, earning Rs 75 per day. After subtracting all costs, the committee of five calculated to be able to make a total profit of about Rs 35,000 by this activity.

3.1.2. Withdrawal of Resources in Muru

Muru is about 6 km in Northern direction from Barwadih near to the Betla wildlife reserved forest. From 1989 to 1994, a plantation was created on about 110 acres of land. The project is continuing on a new plot in a nearby village since January 1994. The president of the *gram samaj* in Muru, Elies Lakra, pooled his complete plot of 20 acres since 1990, but after 3 years of continuous destruction by the local wildlife he withdrew his land from the pool. Whatever was planted was destroyed by herds of wild elephants. These elephants keep on coming from August to December. Social fencing does not work in such cases. Before the planting was started on 10 acres of his land, he had good help in removing stones from his land which became fit for cultivation again. The total plot was subjected to a multitiered cropping pattern. Initially, there was some output from sales of vegetables.

Presently 10 acres of his private land are left barren and function as a buffer zone. When wild elephants reach this buffer zone and there is no agricultural crop in this buffer zone, then the wild elephants return to the wildlife reserved forest, argues Mr Lakra. In this way, the elephants are not attracted to the 10 acres of land under cultivation, which presently yields 5 quintals of foodgrains and 5 quintals of vegetables per acre. Such a reasonable yield is possible due to irrigation by hand.

Mr Lakra believes that *chakriya vikas prenali* can work; albeit not on his land. Looking back at the first three years of *chakriya vikas prenali*, he complained that the two-weekly meetings were only attended by the students to receive their stipends, while the students did not understand the meaning of *chakriya vikas prenali*. Moreover, the students were quite eager in taking loans, but nobody ever repaid them. As a result of poor understanding, the level of protection has been weak in Muru and only a poor yield was realised. The free-grazing practices of the local villagers continues and presently, the pool is degrading again. This all-in-all finally resulted in the breakdown of the protection of his plot.

3.1.3. In the Interior of the Forests: Sindhorwa

Sindhorwa is a small village at a distance of about 12 km in the Southern direction of Barwadih. Sindhorwa is in the interior of the forests. It is connected by a sandy, stony road which is hardly traversable by a two-wheeler. In October 1989 planting started on 90 acres of barren land. The total pool was contributed by one single person: Madan Krishna Barma, a landlord who lives in Daltenganj.

The accumulated village fund, which is distributed as loans, has not been repaid by anybody so far. From a social capital perspective, the meetings proved to be a very useful tool for settling intra village conflicts. The good level of participation is also demonstrated in the sense that everybody in the village has worked in the nursery for a longer or shorter period. They even started borrowing money from the village fund of Barhania (Section 3.1.4.) to employ two people in the nursery of Sindhorwa to maintain and protect their plantation.

Other villagers have started to cut fuelwood from the forest again, to sell it in Barwadih. While the forest around Sindhorwa is degrading, the pool of land is still being preserved.

3.1.4. Barhania: Sustainability within Three Years?

Barhania is about 8 km in the south-western direction of Barwadih, on the main road to Mandal.[5] It is dominated by 65 families of the *Munda* tribe, while there are also 4 Muslim families. Barhania has been one of the most promising 'infant' villages under *chakriya vikas prenali*, which started in October 1994. The seedlings and saplings, which were prepared in two nurseries were planted on a total area of 88 acres, while 4 tanks and 2 check-dams were constructed within one year. It has been amply financed by the Indo-Canadian Cooperation Office (ICCO) with the main purpose to show that self-reliance is possible with *chakriya vikas prenali* within three years. Besides water harvesting structures and a tree-plantation, there are many other activities as well, like employment opportunities for women. A *mahila mandal*, a women's organisation to look after the development of women, has been started in this village, to coordinate a piggery programme and a pressed-plate-making programme.

Lucas Horro, an active student of Barhania, is impressed with the output from the pool in the first year, as the quality of the resources in the pool is improving very fast and the land yielded about 100 kg of grains per acre. He expects higher returns as *chakriya vikas prenali* goes on. He also sees the meetings are educative; they are the 'theoretical' part of *chakriya vikas prenali*. During the meetings, they learn from each other's experiences and they can monitor their work by hearing about the general progress in the meetings.

Albis Soy, another active student of Barhania, is of the opinion that working for *chakriya vikas prenali* does not mean that other jobs have to be sacrificed. It generally only leads to more working days per year. This is especially true for part-time workers.

3.2. Daltenganj Subdivision

3.2.1. Bhusaria: Reduction of Protection

Bhusaria is on the main road to Barwadih, 12 km from Daltenganj. It is a small village dominated by two rather primitive tribals: *Parahia* (22) and *Chero* (15), while a minority of the families belongs to a backward caste (6). Bhusaria belongs to the first five villages that were covered under *chakriya vikas prenali* in 1986. Presently, they have a plantation in the pool of 102 acres, which is contributed by 11 people. About 25 acres of land is pooled by one landlord, Jagdishwar Singh, whom is still actively involved in *chakriya vikas prenali*.

The geographical position of the pool makes it labour-intensive and difficult to protect. The pool is divided by the main road between Daltenganj and Barwadih. Before pooling, it was used as a grazing ground. It has proved to be difficult to change the attitude of the people to stop sending their cattle in the pool.

In 1996, villagers from Bhusaria have started to send their cattle for grazing again. This means that protection of the pool, i.e. social fencing, is weakening in Bhusaria. The former students who used to protect the pool can no longer do so, because they have to look out for temporary jobs outside the village to make a living. Furthermore, the quality of the land in

[5]See also the case study by Choubey and Mehta [32] for more details about Barhania.

the pool is very low, because of many stones. This has reduced the speed of growth of the trees in the pool.

Yet, in 1996 they had about 3500 guava trees which bear fruit. A simple calculation indicates that a single tree can yield more than 10 kg, while the market price is at least Rs 3. Hence, sales of these fruits alone could yield over Rs 1 *lakh* per year. However, sales do not take place. The fruits are either consumed directly by the villagers themselves or stolen by other villagers. For instance, during an interview with two temporarily employed students in the pool through the watershed management scheme, a group of women tried to steal some guavas. After they were spotted, they fled. This shows the difficulty in protecting the pool and obtaining the potential benefits.

Jogeshar Parahiya, an active student, claims that the idea behind *chakriya vikas prenali* works, however, the system of social fencing collapsed due to nonavailability of funds since December 1994. This also resulted in a reduction in the number of meetings from twenty-four to only a few per year. He also narrated that one person gave up working in the nursery, because he found the stipend too low, which was Rs 9 +1 (for the student fund) in 1986.

3.2.2. A Joint Plantation in Tandwa and Sakanpirhi

Tandwa and Sakanpirhi are in the Patan block of Palamau, at a distance of about 30 km North from Daltenganj. Both villages are dominated by the *Brahmin* caste (102), living together with a few *Chero* (6) and *Bunia* (17) tribals, and 21 people belonging to the *Kahar* caste. People who belong to the lower caste own too less land for making a living. They live outside the village for nine months per year, but return during the rainy season to take up jobs in their own village. *Chakriya vikas prenali* in these two joint villages has been initiated in 1988. Presently they have a joint plantation on about 110 acres.[6] During a general body meeting of the *gram samaj* it has been decided to use an alternative sharing mechanism for the land-poolers, namely all profits are to be deposited in the so-called *land-pooler fund*. Hence, the total share of the marketed output of all land-poolers together is put into one account. Only once in every five years, the total amount is paid to the land-poolers equally. This decision of the executive committee was possible, because the shares of the pooled land are quite equal.

In Tandwa, Mr Narain Singh was promoted to the position of programme officer, which never happened before to any other villager in Jharkhand. His father said:

> He was very wise. He was the initiator of the *chakriya vikas prenali* in Tandwa and Sakanpirhi. He organised all the villagers to get the most out of them. He even talked to the district commissioner to get the *chakriya vikas prenali* going in their village. Unfortunately, he died in a car accident in 1992. (Based on interview with Gauri Shankar Singh.)

The pool of land is surrounded by the two villages. This strengthens protection of the pool, reducing disturbance from outside villages. Now, their plantation has the most impressive resource stock of all the villages which were surveyed in this study.

---

[6]Sakanpirhi has a second plantation of 60 acres in Mayapur. This is owned by 6 landowners only. This plantation was undertaken between 1993 and 1994. Presently there is a conflict about actual land-ownership. This area is excluded from the plantation on the joined pool in Sakanpirhi and Tandwa.

Chakriya vikas prenali used to be an interesting process in Tandwa. However, many people lost their interest in the process, due to long delays in the final returns. The earlier planted 30,000 papaya trees, which used to bear some fruits, nearly vanished in 1996, depriving the people of intermediary benefits.

A villager from Tandwa made the following remark in 1996:

> Our land used to yield rice, we have been pooling it for eight years now, while it hardly gave any output. Returns are not expected within the next few years. (Based on interview with Brahmdev Singh.)

Hence, some people of Tandwa are regretting that they pooled their land. In both Tandwa and Sakanpirhi, people pooled relatively fertile agricultural land. This shows the importance of timing the benefits in *chakriya vikas prenali*. Benefits may be delayed, but there is a limit, which has been exceeded for some people already. The positive impact of pooling fertile land is that the growth of trees in the third tier has been quite impressive, which, with a bit more patience, can lead to an impressive yield in the nearby future.

4. Discussion of the Village-Wise Experiences

The economic performance of *chakriya vikas prenali* has been low. In Chapri an 'oasis in the desert' consisting of wasteland has been created, but only about 10% of the original plantation is presently still covered with trees. The belief of the people in Tandwa is breaking down, because the stream of benefits from the pool temporarily came to an end, resulting in a reduction of expectations for future benefits. The promised return on investment is taking far longer than expected.

Yet, *chakriya vikas prenali* has had a number of positive impacts too; for instance, it has solved various conflicts through frequent meetings of the *gram samaj*. Furthermore, a number of people got the opportunity to earn a stipendiary salary by undertaking part-time or full-time work. Another positive point is the agreed-upon sharing mechanism which is unique in India. While the agreed-upon sharing mechanism has not always led to an equal distribution, it has led to a better distribution of benefits than existing government run programmes. Spelling out the exact shares beforehand to the participants has created a positive incentive, which pursuaded villagers to pool land.

Table 8.2 shows the number of trees which have been planted under *chakriya vikas prenali* and have survived over the period 1987–1994. The data are taken from 6 'mature' villages. The area which is still covered with trees amounts to 28.8% of the original area, while 26.6% of the planted trees survived.

There is also a difference between villages from the Barwadih and the Daltenganj subdivision. In the Barwadih subdivision, the villages under *chakriya vikas prenali* are closer to the local office of Society of Hill Resource Management School than the villages in the Daltenganj subdivision. For instance, Khura is bordering to Barwadih, while Lanka, Hendehas, and Chapri are at less than 5 km distance from Barwadih. Sindhorwa is most far off at 12 km. In the Daltenganj subdivision the nearest village is Bhusaria at 12 km, while Kumbhawa is at a distance of about 40 km. Hence, it is easier for the programme officers in Barwadih to assist villages. Moreover, a number of villagers visit the office on their own;

Table 8.2. Survival of trees under *chakriya vikas prenali* in 6 villages.

Village:	Area planted (acres)	Area still covered with trees	Number of: saplings planted	saplings survived	Quality of the pool
Tandwa	105	65	229752	94706	5
Sakanpirhi	103	45	155145	56632	4
Chapri/Karmahi	169	23	239231	106070	4
Khura	166	40	245932	11637	4
Lanka	129	8	200598	1811	6
Sindhorwa	95	40	163426	57135	5
Total	767	221	1234084	327991	
Average		28.8%		26.6%	4.7

Note: Quality of the pool: 4=neither good nor bad, 5=slightly good, 6=quite good.
Source: Sinha [194] cross-checked with my field-survey.

whenever the need arises. Many people from the neighbourhood of Barwadih come to Barwadih during the weekly market-day. Such a close support as in Barwadih is not possible in Daltenganj, as it is too far for the villagers to travel to Daltenganj. Whenever a problem arises, they generally wait for the programme officers to visit their village. Still then project coordinators in Daltenganj have the advantage of direct face-to-face communication with the head-office in Daltenganj. Higher level problems can be resolved more easily in Daltenganj subdivision than in the Barwadih subdivision.

5. Conclusions

There are many restrictions in the way for making the experiment of the *chakriya vikas prenali* really flourish as it should. One of the biggest restrictions is the availability of funds. In a developing country such as India, the availability of funds will under any circumstances be low. The staff of the Society of Hill Resource Management School complain that the funds which become available for implementing *chakriya vikas prenali* are too target-oriented, namely an x number of trees to be planted, a y number of tanks to be constructed, etc. A generous 'free' fund, where the exact expenditure can be decided by the village community itself, is very difficult to obtain.

A major threat to the continuation of *chakriya vikas prenali* is an early exhaustion of funds. Because the stipendiary students do not only work in the nursery; they protect the pool as well. Once the stipendiary students lose their stipendiary income, they have to search for alternative means of income, generally outside the village. This reduces protection of the pool. Then, free grazing is likely to happen again, which possibly deteriorates the condition of the standing trees, and reduces the ongoing benefit stream from the pool. An institutional breakdown is a real possibility in such a case.

There is a high risk involved in a long gestation period of third tier trees, as the possibility of theft, forest fires, drought, diseases, etc. increases. In addition, the marketing of high valued trees and obtaining the right to distribute the benefits is far from obvious. Trees cannot be marketed without formal permission from the government, which is very difficult to obtain. There are typically formal institutional constraints in the way of making *chakriya vikas prenali* flourish. Nevertheless, initiatives like these show the high potential and willingness of local people to collaborate among each other at the village level.

Chapter 9

People Against Deforestation: *Van Panchayats* in Uttaranchal

1. Introduction

Pre-independent people's protests in the form of forest fires led to a formal back up of people's managed communal forests through *van panchayats* in Uttaranchal. More recently and more environmental friendly, forest preservation is enforced through the tree-hugging *chipko* movement. This Chapter analyses the various forms of forest management in Uttaranchal at an institutional level. In addition, the experience at the village level in three different districts is studied. The analysis argues that the level of social capital as emerged through *van panchayats* needs to be preserved.

A large and early experience with forest management by the people is the case of the *van panchayats*. *Van* means forest, while *panchayat* means village assembly. The *van panchayat* is elected by the villagers. The members of the *van panchayats* take decisions on the major management issues for the part of forest which has been assigned to them. This type of forest is popularly referred to as *communal forest*. The institution for managing a communal forest, namely *van panchayat* was backed up legally by the British rulers as early as in 1931. Uttaranchal used to be part of the densily populated North-Western State of Uttar Pradesh and Uttaranchal is also referred to as Uttarakhand. Uttaranchal has been formed officially on 9th of November 2000 with Dehradun the State capital.[1] The former Uttarakhand region of Uttar Pradesh was made up of eight hill districts, out of which five belonged to the earlier region of Garhwal: Dehradun, Tehri Garhwal, Uttarkashi, Pauri and Chamoli. The other three districts belonged to Kumaun: Almora, Nainital and Pithoragarh. Since 2000, there are five more districts in Uttaranchal, namely Haridwar (was not part of Uttarakhand), Rudraprayag (used to belong to Chamoli), Bageshwar (used to belong to Almora), Udham Singh Nagar (used to belong to Nainital), and Champawat (used to belong to Pithoragarh).[2]

[1]See also: http://www.webindia123.com/states/index.htm

[2]These links are derived from the internet by comparing the situation in 1991 (http://www.censusindia.net/up.html) with the situation since 2000 (http://www.mapsofindia.com/maps/uttaranchal/uttaranchal.htm). Hence my fieldwork in Chamoli was actually conducted in the present district of Rudraprayag, while my fieldwork in Almora was actually

Presently, more than 4500 *van panchayats* still exist in Uttaranchal. To look into the current performance of these *van panchayats*, eleven *van panchayats* were selected from three districts, namely Rudraprayag and Pauri, and Bageshwar. These villages were selected after consulation with scientists from the Society for Environmental Communications, New Delhi, who had considerable experience in the region. The three selected districts, on the one hand, have many *van panchayats* and, on the other hand, they are relatively easy accessible. Also, by visiting three distinct districts, various facets of and developments within *van panchayats* could be studied.

A brief account of the Chapter layout is given here. In order to describe forest management practices in Uttaranchal, the evolution of *van panchayats* is put in its historical context (Section 2.). Three alternative practices of managing forests are discussed. The main focus is on management of forests through *van panchayats*. The more recent manifestation of the *chipko* movement in the Rudraprayag and Chamoli districts is also narrated. Section 3. takes the analysis of *van panchayats* to the village level, which is derived from my fieldwork, to discuss how the main principles of forest management perform in specific villages. The merits of the administrative back-up and a more general discussion on how people solved their problems in managing communal forests is presented in Section 4.. This Chapter ends with suggestions for improving the performance of the *van panchayats* (Section 5.).

2. Forest Management in Uttaranchal

In this Section, the changes in property rights of common forest are analysed, to describe the way of managing forests in Uttaranchal in a historical perspective. When considered in a longer time scale (see also Section 3.1.), this is a typical example of the transient nature of property rights. A reclassification of the forests, which emerged out of State support for community-managed forests through *van panchayats*, are dealt with as an evolutionary process. The present rules and regulations for *van panchayats* are discussed in some detail. Finally, I discuss the impact of the *chipko* movement.

2.1. Evolution of *Van Panchayats*

In the nineteenth century, people of Uttaranchal had the rights to almost all forest products. Even into the twentieth century, people still adhered to Trail's *san assi* settlement, under which the village boundaries were demarcated as agreed upon in 1832 (Agarwal, [2]). Based on the following two classes of land, village boundaries were formed.

Nap **land** (measured land) consists of all private land holdings.
Benap **land** (non-measured land) is the complement to all private land holdings.

Except for a few 'old' reserved forests, all common lands could be accessed and used by any villager under this settlement.

In 1883 all *benap* land was converted into state property as a 'district protected forest'. However, people's rights in these forests were still effective until 1910.

conducted in the present district of Bageshwar.

Pant [158] points out the consequences of the British ruler's policy to take away the people's basic rights on forest products. People were denied most rights to forest produce, while they were not able to protect themselves against an increased number of wild animals who were destroying their crops on a large scale. Pant was apt to raise the suspicion that 'the government cares more for the protection and preservation of wild animals than of their human neighbours' [158, page 69]. He suggested that people should be given all rights to hunt animals which intrude upon their private lands. There should be no restrictions on collection of forest products to the extent that no valuable timber is sold. Timber was the main resource that the British rulers needed at that time. Somanathan [196] saw the State as single-minded and self-interested. They stimulated commercial industry to exploit the forest. Existing community-based protection was completely neglected. This eroded the existing norms and values concerning the protection of the forests and valuable social capital was beginning to erode.

Denial of the people's rights in all *benap* lands, led to protests of villagers through large scale fires in many reserved forests of Uttaranchal in the 1920s. By acting in this manner, the people showed to the government that they still had some control on the forests and that they could not be denied entry in these forests without being properly compensated. In response to these protests, the forest grievances committee was set up in 1921 to look into the grievances of people with respect to forest use. They published a report which was accepted by the State administration (see Pandey [157, pages 162–179]). The fires proved effective, because the colonial rulers were forced to revise the policy subsequently. These protests showed that 'persuasion and not coercion can secure people's support in forest protection' (Lal [106]).

Following the suggestions of the forest grievances committee, the reserved forests were reclassified into two groups.

- Class I (non-commercial) reserved forest (oak, etc.): people re-obtained free access. The resources would no longer be exploited by the State.
- Class II (commercial) reserved forest (pine, deodar, etc.): denial of people's rights was maintained in these forests.

The State went as far as to allow *every bonafide resident of Kumaun (and Garhwal)* to exploit class I reserved forests. Hence, a law to dismantle the traditional protection system was formulated to enable the exploitation of the forests. It became an open access regime, where people from neighbouring villages also entered the class I forests of designated villages. This led to heavy degradation of the class I reserved forests.

The villagers themselves thought of a way to deny entrance of other villagers into their forest. Several villages grabbed a part of forest and closed it by appointing guards. In spite of many conflicts, it became a viable way of protecting a part of these forests. People formulated their own rules regarding forest management. This kind of forest management is known as *latha panchayat*. The rules were enforced by the *latha* (stick) carrying guard. This is a typical example of a spontaneously evolved common property regime: a part of forest managed by a well-defined group of users. This voluntary reaction of people against deforestation, prevented the worst effects of 'no management at all' in an open access regime. After 1931, when the law on *van panchayat*s was approved, most *latha panchayat*s were converted into *van panchayat*s. In this manner, the villagers could guarantee a steady supply of primary goods from the forest. The communal forests of the *latha panchayats* are maintained rather well. They still exist in Tehri Garhwal and Uttarkashi, where hardly any *van panchayat* has been created so far.

Except for reversing the policy of the British rulers not much changed after independence, as the government of India simply maintained the strategy of the colonial rulers and even strengthened it (Guha [70, 71]). The major focus of the forest Act was its stress on an idealised State, through which they sought effective forest management. There is a clear incentive to deplete India's forests, because local people are ready to work for extremely low wages, while merchants can gain large profits from timber logging. The yield is far higher than the cost of depletion. The community changed from 'sacred' forest protection into an antipathy against the State. This also contributed to an alienation of the local inhabitants.

2.2. Three Systems of Forest Management

Presently, the forests around the villages can be classified broadly into three groups.

The reserved forest is managed by the forest department. Guards are retained by the forest department to protect the forest.

The civil/soyam forest[3] is under the control of the revenue department. This class of forests is generally not protected.

The communal forest is managed by a *van panchayat*. Guards are retained by the *van panchayat* to protect the forest.

The total geographical area, the total common forest area, the number of *van panchayats*, and the total local fund are shown district-wise for the whole of Uttaranchal in Table 9.1.

Van panchayats were first initiated in the Kumaun region. They are of particular interest here, because they resemble a common property regime. Later, it spread also to two districts of the Garhwal region, but it never reached the Dehradun and Uttarkashi districts and only to a very limited extend the Tehri Garhwal district. It is also interesting to note that Tehri Garhwal, Uttarkashi and Chamoli used to be under the control of a local ruler, the *maharaj*, and were never under the British rule.

Different characteristics of management systems of the three kinds of forests are displayed in Table 9.2. There are many exceptions to the classification in property regimes as shown in Table 9.2. For instance, some commercial forests were not managed at all and resulted in an open access regime, while in some instances reserved forests were grabbed by people and are managed as a common property regime, like a "*latha panchayat*".

The *gram sabha* (village assembly) is, besides the forest *panchayat*, a village level institution which evolved under the *panchayati raj* (village level governing bodies) system. It consists of the president (*pradhan*) and a number of managing members (*panches*) among whom at least two should be female managing members. Civil forests can be used for any purpose as agreed upon by the villagers in the *gram sabha*. It is shown in Figure 9.1) how the *gram sabha* fits into the Indian bureaucracy. There are 6257 *gram sabhas* and 15166 villages in Uttaranchal (UP [205, page 19]). Sometimes, civil/soyam forests are managed voluntarily by the *gram sabha*, but this is generally not the case. That is why the civil/soyam forests can be classified as open access regimes. This means that about 30% of all villages

[3]The civil/soyam forest were earlier called *protected forests*.

Table 9.1. A district-wise overview of forest types and *van panchayats* in Uttaranchal.

District	geographical area (km^2)	common forest area (km^2) total	reserved	civil/ soyam	communal	*van panchayat* number	local fund (Rs 000)
Almora	5385	3888	1439	1821	628	1770	16637
Nainital	6794	4020	3620	192	208	203	2233
Pithoragarh	8856	3241	1317	1210	714	1066	2310
Chamoli	9168	5203	3639	1043	521	668	2445
Pauri	5397	4581	2399	1806	376	924	1030
Uttarkashi	8016	7103	6955	148	0	0	0
Tehri Garhwal	4421	3975	2694	1278	3	14	0
Dehradun	3088	1958	1442	516	0	0	0
Total	51125	33969	23505	8014	2450	4645	24655

Note: The total size of the communal forest was 2418 km^2 and the number of *van panchayats* was 4290 in 1984. (Pandey [157]). Hence, the number of *van panchayats* is still increasing.

Source: Reconstructed from three references: the geographical area is taken from UP [205], the area of reserved and civil/soyam forest are taken from Purohit [165], and the size, fund and number of *van panchayats* are taken from Saxena [178].

Table 9.2. Three kinds of forest management.

Managing body:	*gram sabha*	*van panchayat*	Forest department
President	*Pradhan*	*Sarpanch*	divisional forest officer (DFO)
Kind of forest	Civil/soyam	Communal	Reserved
Quality of forest	Poor	Good	Quite good
% of all forests	23.6	7.2	69.2
Management regime	Open access	Common property	State property

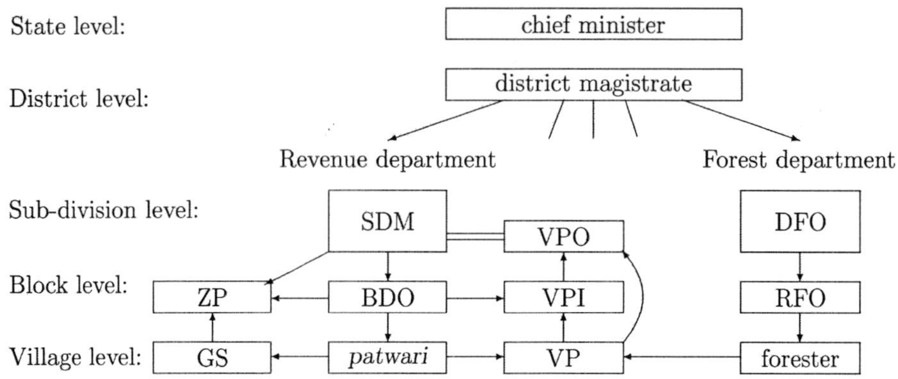

Figure 9.1. Dual control of *van panchayats*.

Note: GS = *gram sabha*, ZP = *zilla parishad* (block council), DFO = divisional forest officer, RFO = range forest officer. For the meaning of other abbreviations, see Figure 9.2.

in Uttarakhand have *van panchayat*s, or halve of all villages within the five districts with *van panchayat*s.

2.3. Main Principles of *Van Panchayats*

A *van panchayat* could be formed, irrespective of the type of forest. For instance, according to Agarwal [2], 'any area within the *san assi* settlement boundary of the village where the residents have rights, recognised by a judicial decision or in a forest settlement,' can be brought under a forest *panchayat*.

The managing body consists of a president (*sarpanch*), and a number of managing members (*panches*). They are elected once every five years, and the *sarpanch* is chosen from among the managing members. During the meetings, which should be held at least four times a year, they decide how to distribute non-timber forest products among the participating members. They formulate their own rules regarding free grazing and stall-feeding of cattle. The following rules and responsibilities exist in *van panchayat*s, since 1976 (Singh [192, pages 241–242]):

a. Protecting and developing their communal forest;
b. Ensuring equal distribution of produce from the communal forest;
c. Fixing and maintaining boundary walls to prevent encroachment;

The main powers of *van panchayat,s* without the interference of the sub-divisional magistrate, are:

A. Determining the sale of dry wood and grass to villagers;
B. In punishing an offender up to Rs 500 (with prior agreement of the offender);
C. In declaring the tools of an offender forfeit;
D. In impounding stray cattle.

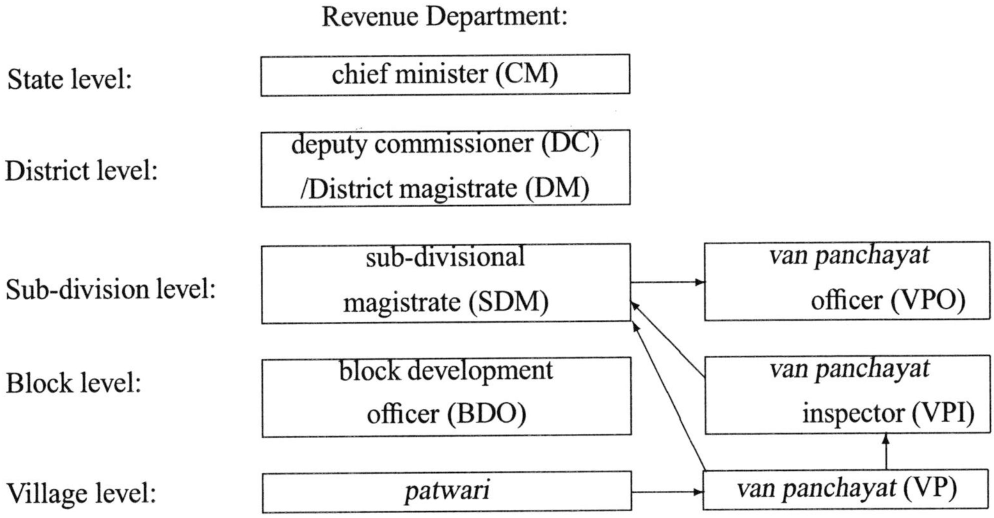

Figure 9.2. The administrative set up for management of communal forests.

Note: The arrows are directed upwards to show through which channels people have to go through before decisions are made.

To protect their communal forest, the managing committee can appoint a guard, known as the *chaukidar*. However, a guard cannot be appointed without prior approval of the sub-divisional magistrate. This guard is generally paid a salary of about Rs 500 per month. This money has to be collected from the members, from door to door. Whenever the people are not ready to contribute sufficient money, a rotational system of guarding is used, where one or two people of the village daily patrol the forest. This system is known as *muwasawari* (Singh and Ballabh, [193]). It is not sure whether, the paid guard or rotational system is more effective for protection of the forest. Britt [24]reported a case where women favoured the rotational system of guarding. The women found a paid guard a waste of money. The rotational system can be done without spending money and it leads to more involvement of people and better protection.

Van panchayats with large communal forests (above 500 ha) generally employ two guards. Guards can also be employed through the *gram sabha*, either to check the civil forest or to protect the private lands against theft or destruction of the crops by grazing or from wildlife. A paid guard is the most important aid in forest protection.

It is a common feature in the *van panchayat* to divide the communal forest in several parts. Some parts are open for fodder and fuelwood collection for a few weeks per year only, while other parts are closed completely. This system avoids a rapid depletion of the communal forest. The period of collection, which differs from resource to resource, is decided during the meetings. This is another strategy for preserving the resource stock.

The administrative set-up for managing communal forests is shown in Figure 9.2. The *van panchayat* inspector, who is appointed by the *van panchayat* officer is very important to keep the process going. However, they are few in number, arguably far too low for effective management.[4] In addition, the *van panchayat* inspector hardly visits villages, because of the lack of financial provisions. The system could become more effective with at least one *van panchayat* inspector at the block level, rather than the actual number of one at the sub-divisional level. The *van panchayat* inspector is a full-time employee concerned with the *van panchayats* at a higher level. The *van panchayat* inspector has many responsibilities like

> conducting elections, monitoring the work of the *van panchayats* and reporting on their functioning to the *van panchayat* officer and the district magistrate. (Singh and Ballabh [193])

Several management problems in the *van panchayat* have to be solved with the help of the sub-divisional magistrate.[5] The sub-divisional magistrate assists the district magistrate who is the person-in-charge at the district level. Until 1956, there was a separate post for a *van panchayat* officer. This task has been taken over by the sub-divisional magistrate. The task of the sub-divisional magistrate is supervisory and advisory. However, the sub-divisional magistrate, who functions as a *van panchayat* officer, has many other responsibilities as well. As a result, guidance of *van panchayats* turns out to be low on his priority list. In the revenue department, the *patwari*, as shown in Figure 9.2, is the actual revenue collector at the village level.

Management of communal forests is administratively backed up by two departments. This so-called dual control of the communal forest is shown in Figure 9.1. The administrative power is vested in the revenue department, while the technical power is vested in the forest department. Before large fines can be imposed, prior approval of the district magistrate is required. Also prior approval of the divisional forest officer is required before timber surplus can be sold. For the commercial sale of trees, prior approval of the district magistrate and the divisional forest officer is required. Hence, the higher the value of the resource, the more bureaucratic hurdles have to be taken.

From 1931 until 1972, people of the villages got 100% of all income generated inside a *van panchayat* through the sales of surplus from their communal forest. This is on and above their own use and consumption of non-timber forest products. In 1972, an amendment in the law on *van panchayats* resulted in reallocation of the surplus income generated through sales of timber and resin. By 1972, the autonomy at the village level reduced, and a system of dual control for managing communal forest was strengthened (Figure 9.1), by introducing a new '2:4:4' sharing mechanism, where the three involved parties, the revenue department, the forest department, and the *van panchayat* would share the sharable surplus as follows:

20% is put aside for the *zilla parishad*, the controlling officer of the *panchayati raj* system at the block level, to plan for managing the communal forests.

The remaining part is shared equally between the Uttar Pradesh forest department and the *van panchayat*.

---

[4] There is only one *van panchayat* inspector for 203 *van panchayats* in Nainital. There are 4 in Almora to look after 1,770 *van panchayats*.

[5] See for example the study on Bachher in Section 3.2.1., where the resignation of the badly functioning *sarpanch* was demanded in front of the sub-divisional magistrate.

40% is deposited in an account for the forest department, to be invested in the improvement of the communal forests. However, the forest department has hardly ever spent any money from that account on the communal forests.

40% is deposited in an account for *van panchayats*. However, it is decided by the *van panchayat* officer or the block development officer where the money should be used for.

Hence, the power of the *van panchayat* was reduced considerably. Their share dropped from 100% to 40% of their share in the surplus of timber produce. The remaining 40% is still controlled by the revenue department. Effectively, this change in rule, deprived the *van panchayat* from any share in the sales of timber surplus, which might erode the level of social capital.

2.4. Women's Involvement in Forest Management: *chipko*

As a reaction to having no role at all in the *van panchayat*, the women started organising themselves as *mahila mangal dals* (women's organisations),[6] which is a managing body at the village level for women only. Such women's organisations, with generally very limited financial resources, were set up to empower women. Besides the president (*adhyaksh*), the organisation has a secretary, a treasurer and a number of managing members. *Mahila mangal dals* in Uttaranchal showed their strength through the *chipko* movement, which arose in Uttaranchal to halt tree-felling (Gadgil and Guha [61]). *Chipko* literally means hugging.[7]

The first *chipko* agitation took place in Mandal Chatti in April 1973, when the Simon company of Allahabad, a sports goods factory, was allotted ten ash trees by the Uttar Pradesh forest department (Kunwar [104]). The same company showed up in Gopeshwar two weeks earlier. When people of Gopeshwar 'learnt of this, they marched to Mandal with drums and songs, gathering more people on the way. A confrontation ensued and the agitators hugged the earmarked trees to foil the attempt of the contractor' (Anil Agarwal in [104, page 19]). This agitation was initiated by men, but women actively participated in it (Qasim [166]). The first agitation by women occurred in Phata Rampur in December 1973. This was initiated by five women: Shama Devi, Indra Devi, Parwati Devi, Jethuli Devi and Jainti Devi.

The real climax of *chipko* occurred in Reni,[8] 25 kms from Joshimath on 26 March 1974. A *mahila mangal dal* was existing in Reni since 1972, with Gaura Devi as the elected president. The forest department had leased an area of about 16 ha with around 10,000 trees for felling, to a contractor. They were afraid to be stopped again as it happened in Mandal Chatti and Phata Rampur a year earlier. So they made a plan to attract all village men from Reni and Lata to Chamoli town, which is about 70 kms far, by promising some work. On the same day, the contractors went with about 60 people in trucks to the forest of Reni.

[6]These women's management bodies have been introduced under different names all over India from 1970 onwards when Indira Gandhi was the prime minister.

[7]The *chipko* movement became known as a protest of women against tree felling by way of embracing a tree. Thus, to cut as tree, one would have to kill the hugger first.

[8]The story of Reni is based on Bist [20] and Kunwar [103]. An account of the *chipko* movement can also be found in Kunwar [104].

When the women came to know about the arrival of these contractors, they stopped working to meet with the contractors. While the women kept on explaining how essential the trees were for them, the contractors were committed to cut the trees. They were wondering how these women would possibly stop them. However, Gaura Devi challenged them by saying:

> Shoot us and our “mother’s house”, but before you can cut any tree you must face us first, if you have the guts, come and kill us. (Kunwar [103])

The protest in this village led the forest department to revise their policy. Their master plan of cutting a total of 13007 ha of forest was finally cancelled (Kunwar [104]). Instead, a process of afforestation was encouraged in Uttaranchal.

There was a big flood on 20 July 1970. This flood originated in the Dhauliganga, Rishiganga and Alakananda rivers. It badly affected the irrigation channel starting from Rishikesh until Roorkee. It killed 55 people and 145 animals (Bist [20]). Much property was destroyed in the Chamoli district. The nearby laying town Gopeshwar, which was not affected by these floods, became the district headquarters in favour of the lower-situated Chamoli town. The floods did not lead to a change in policy at the Indian forest department Moreover, the manifestation of *chipko* in Reni was needed to stress the importance of forests for the people.

The field survey reveals that the four sample villages of the Rudraprayag district have the best average level of resources, $\overline{R} = 0.595$ (see Table B.20 in the appendix). Compare this to $\overline{R} = 0.563$ in Bageshwar and $\overline{R} = 0.329$ in Pauri. While there are many factors which can explain differences in a (perceived) quality of the forest, it is still interesting that the data of the primary survey finds the highest level of resources in the Rudraprayag district. This also suggests that there has been a significant and positive impact of the *chipko* movement, which manifested itself mainly in the Rudraprayag and Chamoli districts.

3. Village-Wise Case Studies

This Section discusses the present situation in the field in the surveyed villages, to see how the principles of *van panchayat*s are practised. The main focus is on *van panchayat*s. The case studies presented below give a flavour of different types of management issues, special characteristics, and conflict resolutions. Figure 9.3 presents a map of the location of the fieldwork undertaken in Uttaranchal. Figures 9.4, and 9.5 provide two illustrations of the situation in the field.

3.1. District Pauri

3.1.1. Dedicated Guard Resigned Honourless in Ulli

Ulli village can be reached by a trek of 3 km from Chaubetta Khal which is 2 km from Khirsu. The reserved forest is shared with a number of other villages. The *van panchayat* became effective between 1950 and 1955. Before 1947, protection of the reserved forests was very effective: nobody could steal wood without being caught. This protection became less strict after independence. This led the people to deplete the reserved forest, simply because it was an easy way to collect resources. One villager explained that, in this village,

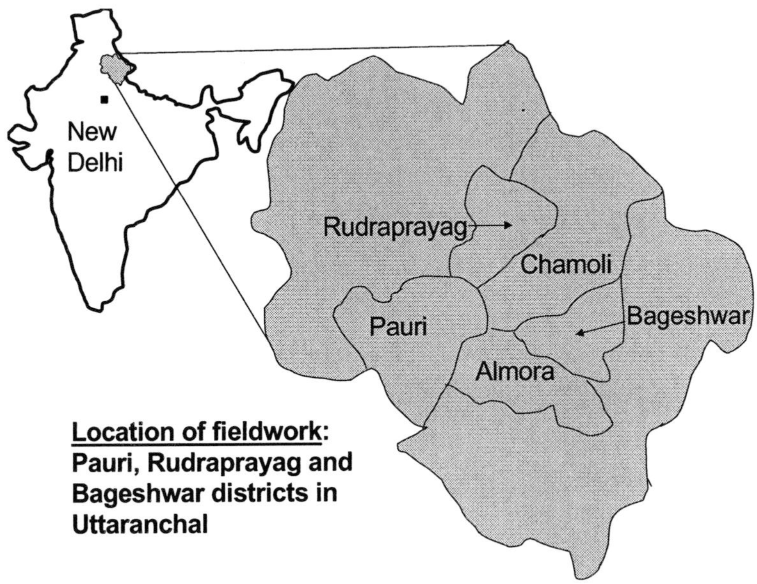

Figure 9.3. Map of districts covered during fieldwork in Uttaranchal.

only 10% of them are presently ready to protect the reserved forest. Obviously, this is not sufficient for sustainable management. However, the negligence of the people towards the reserved forest has not affected their communal forest which is still protected and is in a quite good condition, while the reserved forest is degrading. The forest department already stopped felling trees for timber to reverse the process of degradation.

The reserved forest is open throughout the year. The villagers are not supposed to destroy anything, but it is very easy to deplete. The forest department has employed only one guard who looks after 1500 ha of difficult accessible forest. The villagers collect green leaves from the trees, although this is prohibited by law. However, the wife of the *sarpanch* claims that they cut the green leaves in such a way that the trees are not damaged.

The managing body consists of 9 members, of which 8 belong to the *Rajput* caste[9] and one belongs to a scheduled caste. They hold the minimum required 4 meetings per year in the centre of the village, so that all villagers can attend the meetings.

They used to have a guard to protect the communal forest, but he has grown old and nobody is ready to replace him. To guard the communal forest, daily two persons go to the forest on a rotational basis. A record of who has performed his or her guard duty is regularly maintained. The *sarpanch* and a *panch* sometimes do a *surprise check* to verify whether the two persons on duty are actually doing their job. In spite of that, a number of villagers are still not doing their duty.

Whenever somebody has been caught taking more than their share, (s)he can take the wood, but the axe or sickle is confiscated. If (s)he wants to have the tool back, (s)he has

[9]There are many sub-castes within the *Rajput* caste itself. The caste of a particular person can generally be recognised by the family name. This family name differs from village to village. My sample consists of 84% *Rajputs*. In contrary to the rest of India, a fairly homogeneous caste composition and even distribution of land is very common in Uttaranchal.

Figure 9.4. The communal forest of Shama in the Bageshwar district in Uttaranchal.

Figure 9.5. Village gathering in the thee shop in Ulli in the Pauri district in Uttaranchal.

to pay a fine of Rs 25. These fines are deposited in an account of the *van panchayat*. The communal forest is open from November until January. In that period, one member per family can go to the forest to collect fodder and dry fuelwood.

The forest guard who served the *van panchayat* until 1994 has the following experience to narrate. In 1947 there was hardly any forest in the present communal forest area. The only thing they needed to do to regenerate a forest was to close the area. To raise trees in the communal forest, saplings like pine and deodar were planted. People were taught to cut grass and remove dry wood only. Bigger trees were not to be touched. This was done without any financial help from the government. The guard sometimes caught thieves late at night. He had built up such an experience over the last 40 years that he could recognise immediately when something had been stolen or destroyed in the forest. Some of these thieves could only be caught at night. Once caught, these thieves generally did not come again, so that the communal forest could be regenerated again.

Recently, the guard resigned, because the villagers were not ready to increase his salary. His dedication of 40 years has not been appreciated. It simply resulted in a conflict between the villagers and the guard. The villagers do not value the communal forest high enough to increase the salary of the guard anymore. The guard also has no pension. He claims that about 20 families in this village can easily afford an increase in their financial contribution to employ a guard. But the other 70 families find it a problem, as a result no new guard can be appointed.

The communal forest was of a good quality until last year when major fires occurred. During these fires the guard saw 50% of his work of the last 40 years go up in flames. Since he had already resigned from his post in the *van panchayat*, he could do nothing. After his resignation there has been no interaction between him and the *van panchayat* anymore.

3.1.2. Management of Civil Forest through Women in Than

Than village, which is next to Ulli, is an example where most of the managing members in the *gram sabha* are women. The *pradhan* is also a woman. They are preserving a part of the civil forest of 50 ha very well through the *gram sabha* by employing a guard. The *van panchayat* is hardly worth mentioning, because the size of the communal forest is only 6 ha. In 1995 they got a grant of Rs 50,000 from the forest department to plant trees in their communal forest. The guard is checking both the communal and civil forest. Hence, the actions of the *gram sabha* and the *van panchayat* are coordinated in this village. Meetings of the *van panchayat* are hardly held, whereas the *gram sabha* is called at least 4 times per year. The *van panchayat* is still functioning, because of the *gram sabha*. In this small village, people are ready to pay Rs 180 per family per year to finance the guard for protecting the civil and communal forest. If somebody is found collecting resources against the rules, (s)he has to pay Rs 20 to the *van panchayat* to get his/her sickle back.

The civil forest is open throughout the year for the collection of dry fuelwood, while the communal forest is open only in December and January. Yearly, fodder can be collected for six months from the civil forest and for one month from the communal forest. Hence, there are more restrictions on the communal forest than on the civil forest. In times of scarcity, people can also collect fuelwood from a patch of reserved forest. This forest is difficult to access and it is about one km far. Under normal circumstances, all primary goods are supplied by the civil and communal forest alone.

The people of Than face more difficulties than people of Ulli. If people from Than want to collect goods from the reserved forest, they have to pass Ulli. They are stopped by the villagers from Ulli, who are afraid that the villager's collection accelerates the degradation process of the reserved forest. There is also an extreme shortage of (drinking) water in Than, and while a number of people in Ulli irrigate their land with excess water, people of Than are denied access to this water.

3.1.3. Institutional Breakdown in Thamana

In 1936, a *van panchayat* was created in Thamana village. They used to have 10 ha of communal forest, but it is no longer in use. The communal forest has been handed over to a government official in Pauri in 1990. That forest area is presently controlled by the forest department. In addition, there is also a reserved forest of about 100 ha in the vicinity of Thamana.

This village is a three-km trek from Khola Chorin, which is 14 km from Pauri in the direction of Devprayag. Presently, most of the people collect fuelwood from their private lands. These lands are mainly used for grazing. The amount of land under cultivation is decreasing in this village, because many families left the village, while they did not sell their land.

For a special occasion, timber can be obtained from the reserved forest by writing an application to the divisional forest officer. People who want to cut a tree without permission are fined. In this easily accessible forest, the chance of not getting caught by the 'forest guard' of the forest department is very small.

The people of Thamana do not depend much on the forest. Their level of forest dependence is the lowest of the whole sample: 0.231 (see Table B.20). Depletion of the communal forest started in 1989. Many people started stealing wood from the communal forest and a number of them belonged to one of the four other surrounding villages. Since the guard failed to stop this, the presiding *sarpanch* resigned. The managing committee of the *van panchayat* tried to maintain the communal forest, but the users did not participate. Around 1980, the communal forest was completely fine, while in 1996, only 70 families and 300 people live in this village, while there used to be 125 families and 400 people about 10 years ago. The 'tragedy of open access' in this village already led to an out-migration of 25% of the people.

3.1.4. Biggest Communal Forest of Pauri District in Nishni

Nishni village is a four-km trek from Bhuva Khal which is 6 km from Pauri town, the headquarters of the Pauri district. This village was even a model village during the British rule, when Nishni was the single largest village of Pauri. Since 1996, there are at least two more villages in the district, which are of a similar size. In 1996, there were about 500 families in Nishni as against 60 families in 1901. While there is no civil forest, a *mahila mangal dal*, with 13 managing members and 61 members applied has been registered since 1996.

The villagers do not know the exact size of the communal forest, but the forest is so big that it will take more than 6 hours to walk around the communal forest on the boundary. The size is somewhere between 1,000 and 1,500 ha. The communal forest is open in January and February for fuelwood, but it is open throughout the year for fodder grass. When the communal forest opens, people extract fuelwood very rapidly in order to build up a stock for the whole year. This rapid extraction reduces the quality of the communal forest, but it is better than having free access throughout the year, because in this way the communal forest gets a fair chance to regenerate. Furthermore, it is difficult to protect the communal forest, which is also accessible from other sides by people from other villages. For better protection, they need help from the government to make a boundary wall around the communal forest. There is already a boundary around the communal forest, which is made out of leaves and branches, but this is insufficient to deter illicit entries of non-members effectively.

Encroachment on the forest is another problem in this village. A few years back, one person encroached upon 50 hectares of the communal forest. Initially, nobody could stop him. But after changing the *sarpanch*, the situation improved. The new *sarpanch* took back the encroached land and it became a part of the communal forest again.

There is a scheme from the Uttar Pradesh government to plant trees in the communal forest, once every five years. The saplings are taken from a nursery near Bhuva Khal. They arrive in the village by truck. This provides employment to a number of villagers. They earn Rs 30-35 per day, depending on the type of trees which are being planted. After planting the saplings, the future harvest can be claimed completely by the villagers of Nishni. The government provides only 4 types of trees. Oak is most popular among these villagers, because oak provides fodder and retains water.

M. S. Rawat has been the driving force behind the protection of the communal forest, even before they got the legal status of a *van panchayat* in 1943. Before 1943, the communal forest was already, but unofficially, managed by the villagers. It was said that, M. S. Rawat was so committed to protect the forest, that he would rather have liked to destroy his own house, than to see the forest degrade. He passed away forty years ago.

The World Bank had provided the government of Uttar Pradesh an extensive scheme to develop the region of Uttarakhand. This scheme ran from 1988 until 1993. Under this scheme, schools were constructed, drinking water was provided, improved seeds and fertilisers were provided, etc. The villagers complained that 80% of the scheme's funding has been spent on governance by people in power in the bureaucracy. As a result, intended services to the people did not reach them, also by lack of knowledge of the people about how to get them. For instance, fertilizers, which were freely available, were too difficult to get.

In 1995, officials from the World Bank were supposed to come and inspect the results of the scheme. The following events happened, according to local villagers, in response to only minor improvements in the 5-year period. Some people who were in control of the scheme bribed some villagers, who had no stake in the forest, for putting the forest on fire. In this manner, they could blame the villagers of their negligence towards the forest to save their own face. Therefore, the scheme did not only deny the people their benefits, it even reduced the benefit stream of their communal forest. Hukam Singh, the guard employed by the *gram sabha* to check private lands, pleaded that the whole of Uttarakhand was on fire as a result of the protests of the people against the government who failed to transfer the World Bank grant to local institutions.

3.2. District Rudraprayag

3.2.1. A Communal Forest which is Managed Strictly by Women in Bachher

Bachher is 20-25 km from Gopeshwar town, in the direction of Pokri. The communal forest is divided into six parts, where each part is being closed for regeneration for at least 3 years.

The main nursery of the *dasholi gram swarajya mandal* (the non-governmental organisation, headed by Mr C. P. Bhatt) is located in Bachher. Five people are permanently employed in the nursery, where 50,000 to 100,000 saplings are prepared per year, out of which about 500 villagers are supplied yearly with free tree-saplings. The total size of the nursery is 0.60 ha. Most of the trees are distributed and planted during the monsoon, but a few are also planted in January. The nursery started around 1980 when the *chipko* movement manifested itself in this village.

The first success of women in Bachher[10] was to get electricity. They followed the advice of Mr Chandi Prasad Bhatt to form a *mahila mangal dal* in their village of which Mrs Kalawati Devi became the president.

In 1980 the male *sarpanch* made a deal with a contractor to cut a number trees from the communal forest in Bachher. However, less than 10 trees were actually mature enough for cutting. When she came to know about this, Mrs Kalawati Devi called an emergency meeting of the *mahila mangal dal*, to halt the mismanagement of their communal forest. They did not allow the trucks to enter their village. In response to this event, in front of the *van panchayat* inspector and the sub-divisional magistrate, they insisted on the resignation of their corrupt *sarpanch*. Their request was granted. The *sarpanch* resigned and Mrs. Kalawati Devi, in addition to her position of president of the *mahila mangal dal*, became the new *sarpanch*. The other male managing members refused to be under the command of a female *sarpanch* and they all resigned.

Bachher is the first village where all the members of the *van panchayat* are women.[11] Hence, the *van panchayat* has been managed by the *mahila mangal dal* from that day onwards. The managing committee consists of a secretary, president, treasurer, and nine managing members. Bachher received the *Indira priyadarshni vriksha mitra* (Indira's exposition of tree-friends) award in 1984.

Presently fires are extinguished rapidly by the villagers of Bachher. The *van panchayat* of Bachher is presently so effective that as soon as a fire occurs, all will go to fight the fire, even at night. There have been many fires in 1995. Everybody in the village was finally mobilised to fight these fires. By 1996, the villagers went to the forest as soon as they see some smoke. As a result, no large-scale destruction has taken place in 1996. In addition, the account at the village level consists of Rs 5,000 in the *van panchayat* fund and Rs 7,000 in *mahila mangal dal* fund.

Since there is no input from the forest department to the communal forest, there has not been a single tree given. That gives the villagers all rights in the forest, argues Mrs Kalawati Devi. All matured trees are allocated for private use in the village. People can apply for wood and they will only get it if they really need it and if it is available. So far, the forest

[10]The story of Bachher is partly based on Kunwar [103, pages 15–23] and supplemented with interviews with people of the village.

[11]Mr Chandi Prasad Bhatt claims that there are at least six *van panchayat*s which consist solely of women. For instance, two more female managed *van panchayat*s can be found in Tangsa and Gopeshwar.

department has not been able to extract any resource from their communal forest.

This village has created a rather privileged situation where the government is denied any share in the benefit stream from the communal forest. This is due to the strength of the *mahila mangal dal*. A similar effectiveness is not found in neighbouring villages, where the sub-divisional magistrate is still able to fell trees.

3.2.2. A Huge Communal Forest in Makkumath

Makkumath can be reached by a 22 km long sandy road from Makku Bandh, which is about 15 km from Okhimath. In 1956, 2,200 hectares of reserved forest was handed over to the villagers of Makkumath, to be managed as a communal forest. This forest is presently protected by two guards of Makkumath and one guard of the forest department. There is also a *mahila mangal dal* in this village. The reserved forest is divided into two parts. One part is managed by the *mahila mangal dal* and is improving. The other part is managed by the *van panchayat* and is degrading.

In 1990, two persons were fined Rs 40,000 for illegal sales of timber. In another case, one person from the forest department had ordered to cut about 4,000 trees, employing about forty persons to cut trees in the communal forest. All villagers were mobilised to stop this cutting. By this means, they stopped this so-called wood-mafia in Makkumath.

The Uttar Pradesh tourist department has a master plan to convert the nearby situated Chopta, Bujgali and the Tungnath into tourist spots. Competition for lodging tourists will greatly affect the nearby reserved forest. The villagers of Makkumath oppose the tourist department's master plan. The master plan should be cancelled to maintain the forest, which should, preferably, become a part of the Nandi Devi national park. A tourist spot in Makkumath will only affect their civil forest adversely.

Nandi Kishor Methani, who became the *sarpanch* in 1988, was suspended in 1996, because he refused to put Rs 7,000, earned through the *van panchayat*, into the *van panchayat* account, which 'disappears' at the State level (see Section 2.3.). He did this in protest to the encroachment of the *Garhwal mandal vikas nigam*, representing the Uttar Pradesh tourist department, of two hectares of communal forest without consultation with the people of Makkumath. There should have been four meetings in the village with the *Garhwal mandal vikas nigam*. When finally a meeting was called, none of the managing members showed up.

A group of people in the village is interested in selling timber. They are also the first one who complain about the malfunctioning of the *van panchayat*. This frequent complaining made the villagers think that they are the most concerned with the improvement of the communal forest. This is how the hero of the wood-mafia has been elected as the new *sarpanch*, but he is not managing the communal forest at all. Those people would rather like to see the forest *panchayat* break down, so that they can have full access to the forest. They have applied for electing another *sarpanch*.

3.2.3. Effective Management of Big Communal Forest in Ushara Tala

Ushara Tala is about 12 kms from Okhimath on the main road to Chopta. It can be reached on foot from Makkumath by a trek of 7 kms or by a daily bus which takes a roundabout route of about 19 kms.

The reserved forest, accessible by people of Ushara Tala, consists of three parts. They have two female managing members in the *van panchayat*. People have to pay a fee of Rs 240 per year to obtain the right to get fuelwood from the communal forest. This money is used to employ a single part-time guard who is employed for 6 months during the slack-season, when the danger of theft is the highest. By this means, the guard can undertake agricultural labour as well. A rotational system of guarding is employed for the other six months by two persons per family for the other 6 months in the high season. The *mahila mangal dal* is not very active in this village, as they hardly meet once a year. The civil forest is not degraded in Ushara Tala. The communal forest is even better. These two types of forests are completely managed by the villagers. The reserved forest is also used by other villages, which is the main reason why it is under threat of degradation. The guarding of the forest department is not very effective in this remote region.

3.2.4. Sagar & Gangol: a Multi-village *Van Panchayat* in Conflict

Sagar is 3 km from Gopeshwar on the main road to Mandal. There are about 50 families in Sagar and 3 managing members of the *van panchayat* are from Sagar. Gangol consists of about 120 families and provides 5 managing members for the *van panchayat*. They employ a rotational system of guarding.

The communal forest is divided into four parts, but it is shared between the two villages. There is a conflict, as people from Sagar want to protect the forest, while people from Gangol want to have free access to the forest. The quarrel does not take place during the meetings, but it occurs in the forest. The decisions made during the meetings by the managing members of the *van panchayat* cannot be enforced in the field, since the people of Gangol are not cooperating. People from Gangol claim that they have a right to this forest and they should not be stopped. The *van panchayat* has been suspended since January 1996 and the forest is temporarily not managed at all.

The communal forest was joined in 1965 to simplify the administration of the *van panchayat*s of both villages. It used to be separated before that. The consecutive period of 30 years has been a period of conflicts between the two villages. *Van panchayat*s were elected and broke down. People from Sagar want to separate the communal forest now. People from Gangol are not in favour of such a division, because they will end up with a degraded part of the forest, while Sagar's share is still in a quite good shape.

3.3. District Bageshwar

3.3.1. Shama Institutionalised Stealing from Forest

Shama is situated 30 km from Baradi, the headquarters of the Kapkot block. Baradi is connected by a daily bus from Almora town during the winter season. The Baradi-Shama road is liable to landslides. Shama is known for a well-maintained communal oak forest. Furthermore, there are at least ten more *van panchayat*s near to Shama.

It is forbidden to extract grass and green leaves from the communal forest. Each offender is fined Rs 15 per headload. Rs 10 has to be paid for a headload of fuelwood, while Rs 15 per square feet is fined for a tree. They collect Rs 9,000 per month by fining their members during the peak season. By this means, the 'stealing' of wood has been institutionalised. The *van panchayat* is able to collect about Rs 80,000 per year, which is enough to maintain two guards. They used to have two guards, but one has resigned in 1996. [12] This income of the *van panchayat* is also spent on making a boundary wall around the communal forest and for planting new trees. People, however, need the resources from the forest and see these fines as the actual price of the resource. Furthermore, collection of dry leaves and dry wood is not fined and one segment of the forest is on a rotational basis open for grass cutting from September until November.

The *sarpanch* argues that the Uttar Pradesh government should provide better breeds of animals to their village. The present unproductive animals should be sold. Both men and women plead for a gas connection in their village to prevent a degradation of the forest. When they want to impose a fine on a big offence, like cutting trees without permission, they first have to get permission from the sub-divisional magistrate, which leads to long delays.

The perception of the villagers about their so-called best maintained oak forest differs. Some people in the village blame others of not being preservation-minded. Nomads from the grazer's caste, known as *aniwals*, go higher up in the Himalayas to graze their animals. They come back to Shama during winters, while letting their animals graze in the communal forest. This is seen as the main cause of forest degradation. They are not able to raise a new tree crop, due to the grazing of about 600 goats and sheep, who destroy the new saplings. Children go into the forest and sometimes harm trees, while other children plant them. The connection of Shama by road also has had a negative impact on the quality of the forest.

4. Lessons for Communal Forest Management

In this Section, the main principles of the *van panchayats* are linked with the village-wise experiences, to point out general problems in *van panchayats*. Wherever possible, I try to give suggestions for improving *van panchayats*.

Van panchayats are weak, because of limited formal powers at the village level, as *van panchayats* are not supported legally for punishing big offenses like encroachment and the timber mafia. However, we did find some examples where social action can help to make rule-breakers comply. Like in Nishni, where a dedicated *sarpanch* could foil an attempt of encroachment. And in Makkumath, where collective action of the villagers suppressed the wood mafia. Still then, centralisation of power in the administration at the State level is seen as the main obstacle in the way for efficient management of the communal forest (Shah [185], Singh [192, 193], Ashish *et al* [8], Saxena [178], Somanathan [196], Ali *et al.* [4], Shyamsundar and Kramer [190]).

The most successful *van panchayat* in the sample was in Bachher. Due to a strong commitment of women, the communal forest is managed as a pure common property regime. They obtained more power than they were entitled to have under the official administration. On the other end of the scale, in Thamana, the *van panchayat* vanished, which has already

[12]There are about 170 families in Shama, which means a cost of about Rs 471 per year for the average supply of 27.7 quintals fuelwood. This is about three times lower than the prevailing market price of Rs 51.7 per quintal of fuelwood.

led to 25% out-migration in that village.

To preserve the communal forest, rules for reduced access to communal forests must be formulated and enforced by a paid guard. However, it is not always possible to employ a guard, mainly due to an inability to collect sufficient financial contributions from the villagers. The absence of a guard affects rule enforcement negatively.

A general complaint is that meetings of the *van panchayat* are not held often enough. The counterargument is that people do not show up when there are more meetings. More meetings can lead to better communication among the villagers so that major problems are discussed in time. The survey of this study reveals that 45% of the interviewed people reported that the required four meetings per year were not met in their village.

A *van panchayat* fund is created at the block level from sales of timber surplus, but these funds are not reinvested into the regeneration of the communal forests and allotted to general development programmes. For the *van panchayat*s, it is very difficult to access these funds. Due to the lack of funds, the incentives at the village level to protect and improve the forest have been eroded. Still then, quite some money has been accumulated in these accounts,[13] but it is not within the command of the *sarpanch*. With such a state of affairs, the power of the *sarpanch* is very limited. In response to the management of *van panchayat* accounts, a treasurer needs to be elected from the villagers. (S)he should manage the 40% share-in-surplus *van panchayat* account directly. This is an important incentive to keep the *van panchayat*s going and should not be left out.[14]

The size of the communal forest has no upper limit, which is generally perceived as a drawback of the *van panchayat* policy. However, it can also be considered as a positive and flexible point in the policy, because it is not always possible to split a forest. It is true that it is more difficult to detect rule-breakers when the size of the communal forest is large. In four, out of eleven *van panchayat*s in this study, the communal forest is larger than 500 ha, namely Nishni, Makkumath, Ushara and Shama (see Table B.19). Out of these four villages, Nishni is facing intrusion of people from other villages in their communal forest, leading to difficulties in maintenance of their communal forest. While in Makkumath, the forest department transferred a reserved forest of 2,200 ha to a *van panchayat*. This could be an example for other villages, but it must be remarked that the transfer of reserved forests to a *van panchayat* is especially viable in remote areas. However, when there is such an apparent abundance of trees, the temptation to steal increases. The management committee in Makkumath effectively stopped the wood-mafia. It would have been more difficult for the forest department to stop this, since reserved forests are particularly difficult to protect in remote areas. Ushara did not report any management problems. In Shama a good *van panchayat* account is accumulated by fining people for rule breaking. They earn almost Rs 1 *lakh* per year by that means. This compensates them for the lack of funds within their *van panchayat*.

In a number of the villages which are covered in this study, the supply of necessary,

---

[13] See, for instance, a random sample in the Rudraprayag district of the account in 35 *van panchayat*s in Purohit [165]. He also gives a block-wise overview of *van panchayat* accounts in Rudraprayag and Chamoli districts. Saxena [178, page 61] reports that 'In the Almora district alone, the total account of the *van panchayat* is Rs 1.66 crores (see Table 9.1), but they are hardly permitted to withdraw Rs 3 *lakh*s annually.'

[14] The government of Uttar Pradesh already recognised the difficulties of many *van panchayat*s in 1982. This resulted in many new proposals, but changes were still pending in 1996.

primary goods from the forests is still guaranteed. This is mainly due to the presence of other forests as well. The communal forest only provides enough for all when the forest is of a considerable size. The average size of the communal forest of *van panchayats* of the 11 villages which are covered in this study is 9 times bigger than the total average size of all the *van panchayats* (compare the numbers in Table reftab:area and Table B.19). This suggests that many small communal forests exist that are too small to supply sufficient primary goods.

In contrary to the rest of India, it is remarkable that the women are looking after the agricultural land in Uttaranchal. They collect the fuelwood and fodder from the forest. They have practical knowledge about the needs and importance of the forest. They also know to which extent the forest can be exploited, to maintain a continuous benefit stream of resources from the forest. The main reason for this is that men are employed to a large extent in the service sector in the plains. Nevertheless, men generally still take formal decisions in forest management. The question remains: How can women become further empowered in managing the forest? One possibility is by allowing the existing *mahila mangal dal*s to manage a (different) part of the forest. Another possibility is by reserving seats for women in *van panchayats*. Bist [21] goes as far to suggest that 66% of the seats should be reserved for women. It that case, it will be impossible that men overrule the suggestions made by women. However, in the latter case, it is less likely that any man stays in the managing committee. In Bachher, for instance, all men resigned when Mrs Kalawati Devi became the *sarpanch*. Actually, due to informal constraints, women are caught in a dilemma. They are the collectors of resources and the rule-breakers, while they have no say in the management of the resources (Britt, [23]).Women's involvement in the management of the communal forest is hidden: they may be absent during the meetings, but they compensate for this by making practical decisions 'behind the scenes' (Britt, [24]).Hence, there is a gap between the formulated and the practised rules.

People who have a high dependency on the forest, generally own a little. They are more likely to participate in forest management, because they have high stakes in its preservation. This is an important link between private ownership and forest dependence. Figure 9.6 gives a schematic overview of the linkages between three types of common resources, namely reserved, communal and civil/soyam forests, and private resources in Uttaranchal. The value of the resource is characterised by its benefit stream as pointed out in Figure 9.6.

5. Summary and Conclusions

This Chapter pointed out that *van panchayats* originated from people's claim in restoring their rights in at least a part of the forest. The main principles of managing *van panchayats* were presented, by critically assessing the administrative backup. Agitations through the *chipko* movement ended a period of logging in the Rudraprayag and Chamoli districts. Village-wise experiences were discussed to point out the local conflicts and by which means these conflicts were resolved. It showed the variation in rules for managing the communal forest. Occasionally, people could create a local fund by imposing a kind of tax on resource collection from the communal forest. By this means a guard could be employed to enforce the rules of the *van panchayat*.

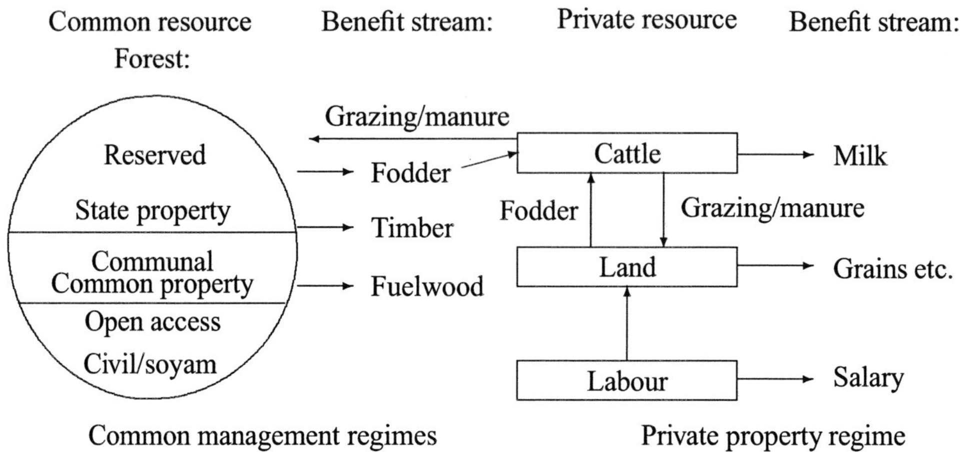

Figure 9.6. Linkages between four types of property regimes.

To improve the functioning of *van panchayat*s, people's original powers should be restored for managing the communal forests. Effectively this means that the managing body of the *van panchayat* is extended by an explicit treasurer to manage the village account to which the 40% share of the *van panchayat* in the sales from the timber produce is transferred directly. Hence, the focus of decision making needs to be at the village level. Then the role of the State should be facilitatory and not directory through coordinating the functioning of several *van panchayat*s and improving the communication between existing *van panchayat*s. In short, the communal forest should be managed by the villagers with participation of the government in people's programme.

Finally, since the women are the main resource collectors, they should somehow be more involved in formulating rules for resource use. This is, however, not easy to establish by a rule change from above, but the trust in women can be improved through dedicated non-governmental organisations.

Chapter 10

Summary and Conclusions

The analysis of this Book aimed at modeling, with the use of econometrics, game theory and institutional analysis, people's participation in forest management, taking cue from three different situations in India. Thereupon, qualitative and quantitative primary data were collected. It has been increasingly recognised that forests can be better managed only when local people are involved. Therefore, Part II of this Book attempted to develop indicators for measuring such involvement or participation. Subsequently, this Book suggested that there is a game among the people who want to gain from using the forest. I developed a method to estimate a game at the institutional level. This game is played between each individual versus other villagers within a number of villages with the same institutional setup. I modeled the voluntary choice for a level of participation in forest management as a repeated game. I have shown that the estimated games can explain whether or not the institution provides incentives for participation in forest management.

The Book has been divided into four parts. Part I served as a general motivation for the study in people's participation in forest management in India. The discussion was at a general level, preparing the reader for the chosen analytical analysis in Part II and the in-depth case studies in Part III. Part IV, which follows after the conclusions, has become a statistical Appendix where the reader can find the developed quantitative questionnaire and an important overview of the gathered primary and secondary data.

In Chapter 1, I gave a short historical account of forest management to motivate the need for people's participation in India. Furthermore, I shortly presented the three case studies and discussed the selection of the case studies at the State, village and household level, together with the database, which is created by undertaking a primary field survey.

Chapter 2 discussed the present trend to give more importance to institutions in economics. This is known as the new Institutional Economics. Institutions can be thought of as a set of formal and informal rules that enable and constrain people's interaction with each other. I concluded that an enforcement mechanism is required to make institutions effective. Management of common pool resources also required different types of institutions depending on property rights, typically fitting into the new Institutional Economics. I also gave an overview of game theoretic approaches to institutions in general and to common pool resources in specific to motivate the use of repeated games. The purpose of this Chapter was to link the specific problem of forest management, which was identified as the problem of managing a common pool resource, to a well established economic framework of analysis.

Part II was analytically most challenging. The quantitative data as collected during the field survey were used for econometric analyses, on the one hand, and for estimating a game model, on the other hand. While econometric analysis of cross-section data has been explained extensively in the literature, an estimation procedure for a game model had to be derived specifically for the type of data used in this study. This estimation procedure was subsequently presented.

Chapter 3 undertook a factor and regression analysis on participatory indicators. The factor analysis showed, in all three situations, that participation in decision making as a societal matter had the highest variation, while contribution to and benefiting from forest preservation had the second highest variation. An econometric analysis of the link between several socio-economic variables and participation in all the three case studies, revealed that better resources, a higher forest dependence, and more equality in the community, enhanced participation.

The factor and multiple regression analysis has been used to address the first research question: 'Which factors enhance people's participation in forest management?' I suggested that the State takes the first move, because they possess resources that local people lack. The process should commence in those villages where participation is most likely to take place, namely among villagers who depend highly on the forest and perceive the quality of the forest as good. The successful village can then serve as an example for other villages to extend the process. Motivated by successes in the first stages, resources can be mobilised to replicate the process in villages with less favourable circumstances. Hence, the process should be an interaction between the State and the people, leading to a win-win situation. Transparency of the State and legal rights for the people are important aspects for success as well.

Chapter 4 provided a taxonomy of two-person symmetric bimatrix games. I presented two methods for estimating a noncooperative game to answer the second research question: 'How can a game model be estimated from primary survey data?' The game model required information about the choice and the gain from that choice, for a number of players. More specifically, members of an organisation formed the players of the game. Through fieldwork, I obtained responses from each villager to several questions related to their participation in forest management. By aggregating or indexing (through a factor analysis) these responses, I derived a level of participation. This is called the choice of the challenger. I derived the choice of a hypothetical second person (contender) from the choices of all other people in an organisation. Information about a number of organisations is needed to derive the game at the institutional level. The Euclidean Cluster Method and the Homogeneous Grouping Method divided the respondents into four groups, respectively based on the Euclidean distances between, or the median of, the choices of the challenger and the contender. With the help of the Homogeneous Grouping Method, the payoffs can also be linked to several socio-economic variables, estimated by a regression model.

In Chapter 5, for each infinitely repeated game, I derived a minimum required or critical discount factor, above which mutual participation can be sustained as an equilibrium outcome. Trigger strategies are used for the chicken's game and the prisoner's dilemma, to derive non-trivial critical discount factors, which have a value between zero and unity. These trigger strategies described a player's reaction to a deviation of one of the players from mutual participation in forest management. The estimated games have shown that all the three institutional settings provide incentives for people's participation in their own way.

I estimated games at the State and village level, in order to address the third research question: 'What can be suggested by application of a repeated game model of peeple's participation in forest management?' By estimating games, I also derived a quantitative measure for the effectiveness and stability of an organisation for forest management. Moreover, I concluded from the estimation results that the studied organisations for forest management in India vary in their effectiveness. Participation was difficult to achieve in Haryana, because of unequal landholding and a great difference between rich and poor. In Jharkhand the villagers were willing to work in the pool, as there were hardly any good labor alternatives. In Uttaranchal the people in the village were quite equitable and were willing to participate when others were willing too. This outcome was based on the assumption that villagers were insensitive to variation in the level of participation of other villagers. The studied organisations for forest management in India turned out to be more stable when villages were sensitive to variation in the level of participation of other villagers in Jharkhand and Uttaranchal. In Haryana, organisational stability could only be guaranteed in the long run, when villagers discounted the future above 0.125.

In Chapter 6, I also considered a finite period penalty, where players defected for a finite number of periods. Furthermore, I derived an intuitive relation between the level of resources and the number of punishments. The analysis showed that, if the payoff structure changes over time, trigger strategies can lose their credibility.

Part III served as a demonstration of the institutional performance in three case studies and addressed the fourth research question: 'What can we learn from institutions that enhance people's participation in forest management?' I linked the case studies with Chapter 2 by studying the different institutional contexts for each situation. When the forest belonged to the State it was called a state property regime, this was the case of joint forest management in Haryana State. When the forest was divisible and where each part was still owned by an individual, while its use pattern was based on group decisions, it was called a group private property regime. This was the case of generating a forest on pooled private land through *chakriya vikas pranali* (cyclic development system) in Jharkhand State. When the forest belonged to a well-defined group of people, it was called a common property regime. This was the case of the management of communal forests through *van panchayat*s in Uttaranchal State. I also considered a fourth category, as a reference point, where forest-ownership was not well-defined. This was called an open access regime. The situation of *no management* was most liable to depletion of the forest.

In Chapter 7, I analysed the evolution and current implementation of joint forest management programme in Haryana. The State has been seeking people's involvement in protecting and regenerating degraded (protected) forests, by giving them more rights on forest products. Since its inception in 1977, it has already led to forty-five registered Hill Resource Management Societies. Successful continuation of joint forest management in Haryana is under pressure, because the people stand to lose their share in the timber from the forest which they protect. It is problematic to share damwater equally because of variation in the landholding pattern. This inequality in landholding can undermine the willingness of landless people to participate in forest management. To compensate these people, every villager has a right to the same amount of water from the dam. Landless people can sell their rights to landowners. Ideally, the mechanism of water rights should work, but it is very difficult to install such a mechanism. In some cases, it was overruled by more powerful landowners who are only interested in the water and indifferent about participation of other people in forest protection.

Chapter 8 studied a process, where people were persuaded, by an outside motivator, to pool land. In some parts of Jharkhand, a multitiered tree-cropping pattern was initiated in the common pool of, generally degraded, private land, using locally available labour. A prenegotiated sharing mechanism, intended to give all parties an equal share, enhanced people's participation. This pooling process, which started in 1986 was operational in more than forty villages in 1996. It became an institution where people were empowered to make their own decisions. However, the actual economic returns were disappointing. This threatened continuation of participation.

In Chapter 9, I analysed the functioning of *van panchayat*s in Uttaranchal. This institution was self-evolving, where people's rights were protected through a registered *van panchayat*, which was approved by law in 1931. The people in turn protected a communal forest. Regeneration of the forest was realised by closing parts of the communal forest and by appointing a paid guard to enforce the management rules as it was decided upon during the meetings. In 1996, more than 4,500 of such *van panchayat*s still existed. Recently, instead of 'enabling' the people, the State reduced people's power in managing their communal forests. This reduced the effectiveness of forest protection through *van panchayat*s.

The main conclusion from a methodological point of view is that this Book provides a manual for undertaking research into the role of people in commonly shared resources. Moreover, the estimated games at the village and the State level do match well with the actual situation in the field. Hence, the estimation method can be recommended for application to any situation where actors have to solve collective action problems in managing a common resources. The main conclusion from an Indian policy point of view is that the studied organisations for forest management vary considerably in their effectiveness and voluntary participation is most likely to emerge among the villagers who depend highly on the forest and perceive the quality of the forest as good.

Finally, I would like to make a few suggestions for possible areas of future research. For instance, similar studies can be undertaken in other States and countries or on other CPR-related issues. This Book has shown that this can be a fruitful exercise. In particular, it would be helpful to develop an econometric foundation for quantifying games to facilitate future applications of game theory. It is my hope that all these endeavours will contribute to a sustainable rural development of future generations.

Part IV

APPENDICES

Appendix A

Quantitative and Qualitative Questionnaires

The answers to each question are recorded by the variable names which are denoted in capitals, either in square brackets or in bold format. This appendix does not only provide the asked questions, it gives a systematic way of recording answers as well. These variables are used for defining socio-economic variables which are used extensively during the data analyses.

1. Household Level Questionnaire for Haryana

1.1. Basic Characteristics to Identify the Household

Name of the interviewed person: [NAME]: [QNUM]
What is the name of your village: [VIL]: [VILNR], [VILGR]
What is your caste: [CASTE]: [CASNR], [CASGR]
What is your religion: [RELIG]: [RELNR]
Have you always lived in this village? (yes/no) If not, when did you come to this village: [VILAR] (year of arrival). How far is that from this village? [VILDI]

1.2. Attitude Towards the Environment

REGEN = To which extent do you consider yourself interested in the regeneration of the forest? ...(1. not interested at all/ 2. not very interested/ 3. somewhat interested/ 4. fairly interested/ 5. very interested)

DEGRA = To which extent do you consider the current state of the forest degraded? ...(1. very degraded/ 2. fairly degraded/ 3. somewhat degraded/ 4. not very degraded/ 5. not degraded at all)

STA10 = What was the state of the forest about 10 years ago? ... Better (1. extremely/ 2. quite/ 3. slightly/ 4. neither/ 5. slightly/ 6. quite/ 7. extremely) Worse

REAS = What do you think are the main reasons for the degradation of the forest? ...(0. Not degraded at all, 1. Negligence by the people, 2. Fires, 3. Overpopulation, 4.

Natural like bad monsoons, 5. Wild animals, 6. Destruction by some people, 7. Negligence by the government, 8. Conflicts in the community, 9. Other reason)

FODQU = What is the quality of the resources, like fodder and fuelwood extracted from the forest? ... Bad (1. extremely/ 2. quite/ 3. slightly/ 4. neither/ 5. slightly/ 6. quite/ 7. extremely) Good

HRMS = If you are a member of the Hill Resource Management Society, then what is your position? ... (0. None. 1. Member, 2. Managing member, 3. Guard, 4. Casher, 5. Secretary, 6. President)

1.3. Indicators of Participation

Respond to the following questions in the following manner:

Not at all ________ little bit ________ medium ________ quite a lot ________ very much
1 ________ 2 ________ 3 ________ 4 ________ 5

Contributing to forest conservation:

REPL = Are you undertaking tree-plantation work?

CONTR = Are you contributing something like money or voluntary labour to the Hill Resource Management Society to help in its functioning?

Sharing in the benefits:

BENEF = To which extent are you benefiting from the presence of the Hill Resource Management Society in your village?

Involvement in decision-making and evaluation through meetings:

FREQ = How frequent are the meetings of the managing committee?

ATTEN = How frequent do you attend the meetings?

FOLOW = To which extent can you agree with the decisions of the managing committee?

INFL = To which extent can you influence the decisions of the managing committee?

HRINT = To which extent do you find the discussions in the meetings of interest?

HRUSE = To which extent do you gain from such meetings?

SUGG = To which extent do you make suggestions during such meetings?

OWNER = Who is the owner of the forest? ... (1. the government, 2. the society, 3. both the government and the society: joint-venture, 4. none of them)

1.4. Agriculture

Land-ownership: Do you own any land? ('0' means no land)

LANDO = How much land do you own? (in acres)

LANDC = How much land do you use for cultivation? (in acres)

LANDI = How much land is irrigated? (in acres)

CANAL = Is your land irrigated by a channel? (yes/no)

SOURC = What is your source of irrigation?

LANDR = Did you receive any rent by leasing your land? (Rs per year)

Table A.1. Outputs from agricultural production.

Total (in quintals)		Amount sold	Price (Rs)	Total yield
1. Wheat	[WHEAT]	[WHEAS] ×	[WHEAP]+	
2. Maize	[MAIZE]	[MAIZS] ×	[MAIZP] +	
3. Rice	[RICE]	[RICES] ×	[RICEP] +	
4. Dhal	[DHAL]	[DHALS] ×	[DHALP]+	
5. Jaggery	[GUR]	[GURS] ×	[GURP] +	
6. Oil	[MUST]	[MUSTS] ×	[MUSTP]+	
7. Vegetables	[VEG]	[VEGS] ×	[VEGP] +	
8. Fruit	[FRUIT]	[FRUIS] ×	[FRUIP] +	
9. Cotton	[COTT]	[COTTS] ×	[COTTP] +	
10. Peanuts	[PEAN]	[PEANS] ×	[PEANP] +	
11. Residuals		[RESIS] ×	[RESIP] =	
Total:	[LANQU]			[LANYI]

FRAG = In how many fragments is your land divided? (#)[1]
LANDS = For how much have you sold any land over the last year? (Rs per year)

Table A.1 presents the outputs from agriculture, while Table A.2 presents the inputs into agriculture.

1.5. Livestock Rearing

Table A.3 gives an overview of livestock ownership and trade, while Table A.4 presents the outputs from livestock.

1.6. Fodder, Fuelwood, *Bhabbar*, Bamboo and Loans

GRAZE = Is your cattle sent for grazing? (yes/no)
GRAPL = Where is your cattle sent for grazing?
GRMON = For how many months does your cattle graze?

Table A.5 presents the fodder purchases from the market, Table A.6 shows the sources of fodder, fuelwood, *bhabbar*, bamboo and loans and Table A.7 gives the sources of loans.

LREAS = Why did you take a loan?

1.7. The Configuration of the Family

The following information is collected about each family member by using Table A.8.

QNUM = Questionnaire number.

[1] Symbol '#' is used to represent number.

Table A.2. Inputs for agricultural production.

	Total input	Price (Rs/unit)	Expenditure
1. Seeds (qu)	×	+	[SEED]
2. Uria (qu)	[URIA] ×	[URIAP] +	
3. D.A.P.(qu)	[DAP] ×	[DAPP] +	
4. Tractor (h)	[TRACH] ×	[TRACP] +	
5. Labour (days)	[LABOH] ×	[LABOP] +	
Irrigation: (h)	×	+	
6. Dam-gravity	[DAMWA]×	[DAMWP] +	
7. Dam-engine	[DAMEN] ×	[DAMEP] +	
8. Tubewell	[TUBEW] ×	[TUBEP] =	
Total:			[LANCO]

Table A.3. Livestock ownership and trade.

Cattle-kind:	Number	Sales(Rs/year)	Purchases
1. ox	[OX]		
2. cows	[COW]		
3. buffaloes	[BUFF]		
4. goats	[GOAT]		
5. chicken	[CHICK]		
6. mule	[MULE]		
7. camel	[CAMEL]		
Total		[ANSOL]	[ANPUR]

Table A.4. Outputs from livestock rearing.

Total (in quintals)		Amount sold	Price (Rs)	Total yield
1. Milk	[MILKY]	[MILKS] ×	[MILKP] +	
2. Cowdung	[TPD]	[TPDS] ×	[TPDP] =	
Total:				[ANIYI]

Table A.5. Fodder purchases from market.

	Amount (qu/year)	Price (Rs/qu)	Expenditure
Fodder grass	[FODMAR]×	[FODMPR] +	
Collection fee		+	[DATI]
Feed	[FEED] ×	[FEEDP] +	
Oil cake (*khal*)	[KHAL] ×	[KHALP] =	
Gross total:	[FODPUR]		[ANICO]

Table A.6. Sources of fodder, fuelwood, *bhabbar* and bamboo.

From:	Fodder (qu)	Fuelwood (qu)	*Bhabbar* (kg)	Bamboo (#)
Forest	[FODFOR]	[FWFOR]	[BABFOR]	[BAMFOR]
Panchayat land	[FODPAN]	[FWPAN]		
Private Land	[FODPRV]	[FWPRV]	[BABPRV]	
Market	[FODPUR]	[FWMAR]	[BABMAR]	
Landlord's land				[BAMLL]
Total use	[FODTOT]	[FWTOT]	[BABTOT]	[BAMTOT]
Market price	[FODPR]	[FWPR]	[BABPR]	[BAMPR]
Society price	[DATI]			

Table A.7. Sources of loans.

From:	Initial loan	Taken on	Present debt	Interest rate
Bank	[LTAKE]	[LDATE]	[LOANB]	[INTER]
Landlords			[LOANL]	
Shopkeepers			[LOANS]	
Total			[LOANT]	

Table A.8. Information about each family member.

Name	Sex	Age	Education	Relation	Employment	Salary	Remittance
1							
2							
3							
4							
5							
6							
7							
8							
9							
10							
11							
12							
13							
14							
15							

FNR = Family member number.

NAME = Names of all family members.

SEX = Gender ... (male=0, female=1)

AGE = Age.

EDU = Education. (years of schooling)

REL = Relation to interviewee. ... (1. oneself/ 2. husband/ 3. wife/ 4. son/ 5. daughter-in-law/ 6. grand-son/ 7. grand-daughter/ etc.)[2]

EMP1, EMP2 = Main employment or activity: ... (1. agriculturist/ 2. animal husbandry/ 3. mixed agriculturist/ 4. share-cropper/ 5. tree-planter/ 6. basket-maker/ 7. temporary worker/ 8. shopkeeper/ 9. teacher/ 10. government servant/ 11. mechanic/ 12. carpenter/ 13. artisan/ 14. industrial worker/ 15. doctor/ 16. truck driver/ 17. washerman/ 18. business / 19. student in government school/ 20. household activities/ 21. guard/ 22. in training/ 23. student in private school/ 24. painter/ 25. hairdresser/ 26. military/ 27. electrician/ 28. pension)

SAL = Total salary from work. (Rs per year)

REM = Total remittances sent to family-member. (Rs per year)

EMPPL = Employment place.

1.8. Consumption

Table A.9 presents the collected data on household consumption.

[2]There are more than 50 different words in Hindi to define family relations.

Table A.9. Amount and value of various private consumption items.

Food items (per year)	Amount purchase	Purchase price	Total purchase	Sub-totals
1. Cereals (qu)				
a. Wheat	[WHPUR]×	[WHPPR]+		
b. Maize	[MAPUR]×	[MAPPR]+		
c. Rice	[RIPUR] ×	[RIPPR] =		[CCER]
2. Non-cereals (kg)				
a. Pulses	[DHPUR] ×	[DHPPR] +		
b. Vegetables	[VEPUR] ×	[VEPPR] +		
c. Fruits	[FRPUR] ×	[FRPPR] +		
d. Sugar/jaggery	[SUPUR] ×	[SUPPR] +		
e. Spices		+	[SPICE]	
f. Tea	[TEPUR] ×	[TEPPR] =		[CNCER]
3. Cattle products (kg)				
a. Milk	[MIPUR] ×	[MIPPR] +		
b. Oil	[OIPUR] ×	[OIPPR] +		
c. Ghee	[GHPUR] ×	[GHPPR]+		
d. Meat	[MEPUR] ×	[MEPPR]=		[CANIM]
4. Intoxicants				
a. *Bidi*			[BIDI] +	
b. Cigarettes			[SIGAR] +	
c. Liquors			[LIQOR] +	
d. Hooka			[HOOKA]+	
e. Chew tobacco			[CHTOB] +	
f. *Pan*			[PAN] =	[CINTX]
5. Energy				
a. Kerosine (ltr)	[KEROS] ×	[KEROP] +		
b. Electricity		+	[ELEC]	
c. Phone		+	[PHONC]	
d. Petrol		+	[PETR]	
e. Gas (tanks)	[GAS] ×	[GASP] +		
f. Fuelwood (qu)	[FWMAR]×	[FWPR] =		[CENER]
6. Miscellaneous				
a. Clothes			[CLOTH] +	
b. Shoes			[FOOTW]+	
c. Medicines			[MEDIC] +	
d. Transport			[TRAV] +	
e. Education			[EDUC] +	
f. Soap			[SOAP] +	
g. Servant			[SERV] +	
h. House rent			[HOUSR]=	[CMISC]
Grand total:				[CONS]

1.9. Capital

Table A.10 presents the household ownership of capital goods and its value.

1.10. Definitions of the Derived Variables

NRCH = Number of children. (#)

NRAD = Number of adults. (#)

NRWOM = Number of women. (#)

NFM = Number of family members: [NRCH] + [NRAD]. (#)

FMRAT = Female-male ratio: 2000×[NRWOM]/[NFM].

NRGSC = Number of children in government school. (#)

NRPSC = Number of children in private school. (#)

NRSGE = Number of children expected to go to school. (#)

NRWFM = Number of working family members. (#)

WFMOV = Number of family members working outside the village. (#)

AVAGE = Average age.

EDU7 = Average years of schooling of those family members who are older than 7.

EDU15 = Average years of schooling of those family members who are older than 15.

TSAL = Total salary earned by all family members. (Rs per year)

TREM = Total remittance sent to the family. (Rs per year)

SEXS = The gender of the interviewed person … (male=0, female=1)

AGES = Age of the interviewed person.

AGEGR = Age group of the interviewed person … (0:0–10; 1:10–20; 2:21–30; 3:31–40; 4:41–50; 5:51–60; 6:61–100).

EDUS = Education of the interviewed person.

EDUGR = Education group of the interviewed person … (0:0; 1:1–4; 2:5–7; 3:8–9; 4:10–11; 5:12–14; 6:15–17).

ENVQU = Environmental quality: [DEGRA] + [STA10] + [FODQU].

PARTI = Level of participation based on the total sum: [REPL] + [CONTR] + [BENEF] + [FREQ] + [ATTEN] + [FOLOW] + [INFL] + [HRINT] + [HRUSE] + [SUGG].

FERT = Total use of fertilisers: [URIA] + [DAP]. (quintals per year)

DAMH = Amount of damwater received: [DAMWA] + [DAMEN]. (hours per year)

ANIMUN = Standard animal unit: [OX] + [COW] + 1.2 × [BUFF] + 0.25 × [GOAT] + 0.10 × [PIG] + 2.67 × [MULE] + 2.67 × [CAMEL] + 2.67 × [HORSE]. (cow-units)

LANIN = Income from land: [LANYI] - [LANCO] + [LANDS]. (Rs per year)

ANINC = Income from animal husbandry: [ANIYI] – [ANICO] + [ANSOL] – [ANPUR]. (Rs per year)

INCOME = Total household income: [TSAL] + [TREM] + [LANIN] + [ANINC]. (Rs per year)

INCPC = Per capita income: [INCOME]/[NFM]. (Rs per year)

CONPC = Per capita private consumption: [CONS]/[NFM]. (Rs per year)

Table A.10. Amount and value of privately owned capital goods.

	Capital-goods:	Number/amount	Value	Total value	Sub-total
1.	<u>House</u>		+	[HOUSE]	
a.	*charpai*	[CHARP]	× [CHARV] +		
b.	bed	[BED]	× [BEDV] +		
c.	chair	[CHAIR]	× [CHAIV] +		
d.	bench	[BENCH]	× [BENCV] +		
e.	table	[TABLE]	× [TABLV] +		
f.	television	[TV]	× [TVV] +		
g.	radio	[RADIO]	× [RADIV] +		
h.	tape-recorder	[TAPE]	× [TAPEV] +		
i.	cycle	[CYCLE]	× [CYCLV] +		
j.	scooter	[SCOOT]	× [SCOOV] +		
k.	motor	[MOTOR]	× [MOTOV] +		
l.	electric fitting	[ELECF]	× [ELECV] +		
m.	gas-cylinder	[GASC]	× [GASV] +		
n.	solar	[SOLAR]	× [SOLAV] +		
o.	fan	[FAN]	× [FANV] +		
p.	refrigerator	[FRIDG]	× [FRIDV] +		
q.	cooler	[COOL]	× [COOLV] +		
r.	phone	[PHONE]	× [PHONV] +		
s.	Shop goods		+	[SHOPG]	
t.	car	[CAR]	× [CARV] +		
u.	truck	[TRUCK]	× [TRUCV] =		[CHOUS]
2.	<u>Land and cattle</u>:				
a.	land	[LANDO]	× [LANDV] +		
b.	oxen	[OX]	× [OXV] +		
c.	cows	[COW]	× [COWV] +		
d.	buffaloes	[BUFF]	× [BUFFV] +		
e.	goats	[GOAT]	× [GOATV] +		
f.	chicken	[CHICK]	× [CHICV] +		
g.	mules	[MULE]	× [MULEV] +		
h.	camels	[CAMEL]	× [CAMEV] =		[CANLA]
3.	<u>Irrigation</u>: tubewells	[TUBEO]	× [TUBEV] =		[CIRRI]
4.	<u>Agricultural implements</u>:				
a.	plough and leveller	[PLOUG]	× [PLOUV] +		
b.	sickle and axe		+	[AXEGD]	
c.	ox cart	[OXCAR]	× [OXCAV] +		
d.	cane-crusher	[CANEC]	× [CANEV] +		
e.	manual fodder cutter	[FCMAN]	× [FCMAV] +		
f.	electric fodder cutter	[FCELE]	× [FCELV] +		
g.	diesel fodder cutter	[FCDIE]	× [FCDIV] +		
h.	thresher	[TRESH]	× [TRESV] +		
i.	tractor+implements	[TRACT]	× [TRACV] =		[CGOOD]

CAPIT = Total value of capital goods: [CHOUS] + [CANLA] + [CIRRI] + [CGOOD] − [LOANT]. (Rs)

CAPPC = Per capita capital: [CAPIT]/[NFM]. (Rs)

PAYOF = Total payoff: 118.0 × [FODFOR] + 109.5 × [FWFOR] + 196.9 × [DAMH] + 3.5 × [BABFOR] + 3.5 × [BAMFOR]. (Rs per year)

PAYPC = Per capita payoff: [PAYOF]/[NFM]. (Rs per year)

1.11. Definition of Forest Dependence

Forest dependence is defined as the amount of goods collected from the forest as compared to the total consumption of these goods, like fuelwood, fodder, *bhabbar*, and bamboo.

$$\begin{aligned}
&\text{Let x} \in \{\text{FW, FOD, BAB, BAM}\}\\
&[\text{xDEP}] = \begin{cases} \frac{[\text{xFOR}]}{[\text{xTOT}]} & \text{if xTOT} > 0 \\ 0 & \text{if xTOT} = 0 \end{cases}\\
&[\text{xNR}] = \begin{cases} 1 & \text{if xTOT} > 0 \\ 0 & \text{if xTOT} = 0 \end{cases}\\
&[\text{FORDEP}] =\\
&\frac{[\text{FWDEP}] + [\text{FODDEP}] + [\text{BABDEP}] + [\text{BAMDEP}]}{[\text{FWNR}] + [\text{FODNR}] + [\text{BABNR}] + [\text{BAMNR}]}
\end{aligned} \tag{A.1}$$

1.12. Definition of Gini Coefficient

Income inequality, measured by the Gini coefficient, can be calculated by a numeric approximation of the Lorentz curve.[3] This can be done in five steps as it is shown in formulas A.2–A.6. The approximation at the village level is illustrated in Figure A.1:

$$\text{Let: INCPC}_i \geq \text{INCPC}_{i-1} \text{ for all } i \in N; \tag{A.2}$$

$$\text{TOTINC}_j = \sum_{i \in N_j} \text{INCPC}_i \tag{A.3}$$

$$A_{k_j} = \frac{\text{INCPC}_k}{\text{TOTINC}_j} \frac{1}{2|N_j|}; \quad k \in N_j, \quad j \in V \tag{A.4}$$

$$B_{k_j} = \sum_{i=1}^{k-1} \frac{\text{INCPC}_i}{\text{TOTINC}_j} \frac{1}{|N_j|}; \quad i,k \in N_j, \quad j \in V \tag{A.5}$$

$$\text{GINI}_j = \frac{1}{2} - \sum_{k=1}^{|N_j|} (A_{k_j} + B_{k_j}) \tag{A.6}$$

Here INCPC resembles per capita income (Rs per year), while N is the set of households in the whole sample, N_j is the set of households in village j, $|N_j|$ is the number of households in village j, and V is the set of villages covered in the study.

[3]This was programmed in Turbo Pascal.

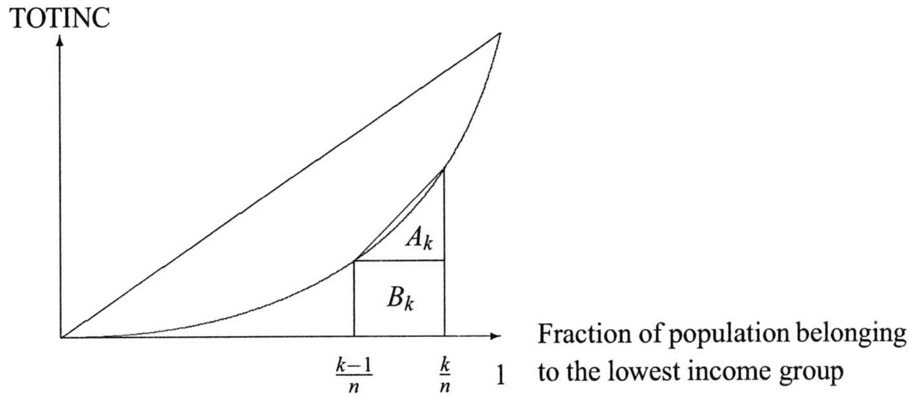

Figure A.1. Approximation of the Gini coefficient.

1.13. Definition of Equality Index

An equality index is developed to measure the homogeneity in caste and religion. This is shown in equation A.7.

$$\text{Equality} = \sum_{i \in K_j} \left(\frac{c_{ij}}{\sum_{i \in K_j} c_{ij}} \right)^2 \quad \text{for all } j \in V. \tag{A.7}$$

Where c_{ij} is the number of people in a particular caste or religion group i in village j. K_j is the set of caste or religion groups in village j. V is again the set of villages covered in this study. The fractions are squared, so that villages with a greater spread in caste or religion have a lower index. The total *equality index* (EI) is the average of equality of caste (EC) and religion (ER).

$$\text{Equality index (EI)} = \frac{\text{EC} + \text{ER}}{2}. \tag{A.8}$$

2. Village Level Questionnaire for Haryana

Who has which function in the Hill Resource Management Society?

1. president: [PRES];
2. secretary: [SECR];
3. cashier [CASH];
4. managing members: [MM1], [MM2], [MM3], [MM4], [MM5], [MM6], [MM7], [MM8];
5. guard: [GUARD];

ELEC = What was the date of the last election?

FOREST, FORQU = What is the size and quality of the forest?
Very degraded 1 ...... 2 ...... 3 ...... 4 ...... 5 ...... 6 ...... 7 very dense

FORDI = Where is the forest situated? How far must the villagers go to collect their resources?

FORDE = How many communities depend on the same forest?

COORD = Is there any coordination between depending communities?

GRAZE = Is free grazing happening?

GRNOT = What is done against free-grazing? Which changes have taken place within the practice of free-grazing? Which actions are taken towards depleters and illegal tree-cutters? How are they caught? How are they punished?

ENCRO = Has anybody encroached upon the forest land? If yes, how many acres?

PROBL = What are the main problems of your society?

Table A.11 shows indicators at the village level.

3. Household Level Questionnaire for Jharkhand

The quantitative questionnaire for *chakriya vikas pranali* in Jharkhand is quite different from the questionnaire for Haryana. People in Jharkhand pool land, which is treated as the common pool resource. They hardly obtain any fuelwood, fodder, and timber from reserved forests.

3.1. Basic Characteristics to Identify the Household

The set of questions used here is exactly the same as in Section 1.1..

3.2. Attitude Towards the Environment

In this Section all questions are adjusted.

REGEN = In the system of cyclic development, people are pooling land voluntarily into a common pool. To which extent do you consider yourself interested in the regeneration of the pool of land? ...(1. not interested at all/ 2. not very interested/ 3. somewhat interested/ 4. fairly interested/ 5. very interested)

DEGRA = To which extent do you consider the current state of the pool of land degraded? ...(1. very degraded/ 2. fairly degraded/ 3. somewhat degraded/ 4. not very degraded/ 5. not degraded at all)

STA10 = What was the state of the pool of land about 2 to 10 years ago? ...(1. much better, 2. better, 3. same, 4. worse, 5. much worse)

FODQU = What is the quality of the resources, like fodder, fuelwood, vegetables and fruits which you extract from the pool? ...Bad (1. extremely/ 2. quite/ 3. slightly/ 4. neither/ 5. slightly/ 6. quite/ 7. extremely) Good

ECMEM = If you are a member of the *gram samaj*, then what is your position? ...(0. None, 1. Member, 2. Managing member, 3. Casher, 4. Secretary, 5. President)

Table A.11. Indicators at the village level

Demographics:		Caste composition:	
population:	[POPUL];	1. *Pandit*:	[PANDIT];
female population:	[FEMPOP];	2. *Rajput*:	[RAJPUT];
# families:	[SAMPLE];	2. *Jat*:	[JAT];
# working family members:	[NWFM];	3. *Gujjar*:	[GUJJAR];
# working people outside village:	[WFMOV];	4. *Harijan*:	[HARIJAN];
		4. Others:	[OTHER];

Amount and kind of land:		Number of animals:	
Owned private land:	[LANDO];	Oxen:	[OX];
Cultivated private land:	[LANDC];	Buffaloes:	[BUFF];
Irrigated land:	[LANDI];	Cows:	[COW];
Forest area:	[FOREST];	Goats:	[GOAT];
Communal land:	[COMMON];	Chicken:	[CHICK];
Number of landholding families:	[NLOF];	Horses/mules/camels:	[DRAUG];

Capital:		Latrines:	[LATRI];
Houses:	[HOUSE];	Bio-gas:	[BIOGAS];
Televisions:	[TV];	Improved *chulhas*:	[IMPCH];
Cycles:	[CYCLE];	Cane crushers:	[CANEC];
Scooters:	[SCOOTER];	Threshers:	[TRESH];
Motors:	[MOTOR];	Fodder cutters:	
Auto:	[AUTO];	manual:	[FCMAN];
Trucks:	[TRUCK];	electric:	[FCEL];
Gas-cylinders:	[GASCY];	diesel:	[FCDIE];
Electricity fittings:	[ELECF];	Rope-making machine:	[ROPE];
Phones:	[PHONE];	Engines (electr/diesel):	[ENGIN];
Solars:	[SOLAR];	Tractors:	[TRACT];

Common resources:	
Dams:	[DAM];
Storage capacity of the dams:	[CATCH];
Amount of land irrigated by the dams:	[COMMA];

School	Primary(1-5)	Middle(6-8)	Matric(9-10)	Inter(11-12)	BA(13-)
Number					
Distance					
Place					

3.2.1. Pooling of Resources

Do you own land? Out of this, how much land have you pooled? [PLAND] (acres) for [PLANY] (years). Are you planning to pool some land in future? [PLEXP] (acres)

Are you also doing any kind of job with respect to *chakriya vikas pranali*? [PJOB] (days per year), for [PJOBY] (years).

3.2.2. Availability of Forest Products

FUSE = Are you using any products from the forest like fuelwood, fodder, leaves, bamboo?

FUFAR = Do you have to go far to collect these products?

FGRAZ = Are you able to let your cattle graze free in the forest?

3.3. Indicators of Participation

In this Section the participatory responses are changed considerably. The following response-pattern is used again:

Not at all ______ little bit ______ medium ______ quite a lot ______ very much
1 ______ 2 ______ 3 ______ 4 ______ 5

Contribution to the resource pool:

PEXT = To which extent are you pooling land or labour?

Benefits from the pool:

PBEN = To which extent are you benefitting from the pooling process?

PUSE = To which extent are you able to collect resources from the pool like fuelwood, fodder, bamboo, leaves, fruits, water, vegetables?

PUBEN = To which extent are you benefitting by using the pool of resources?

Involvement in decision-making through meetings:

GMEET = How frequent are the meetings of the executive committee?

GSINP = How important do you consider the meetings for proper functioning of the system of cyclic development?

GPRES = How frequent do you attend the meetings?

GSFOL = To which extent can you agree with the decisions of the executive committee?

GSINF = To which extent can you influence the decisions of the executive committee?

GSINT = To which extent do you find the discussions in the meetings of interest?

GSUSE To which extent do you gain from such meetings?

GSSUG To which extent do you make suggestions during such meetings?

Table A.12. Additional outputs from agricultural production.

	Total (in quintals)	Amount sold	Price (Rs)	Total yield
12. *Mahua*	[MAHUA]	[MAHUS]	× [MAHUP]+	
13. *Chapra*	[CAPRA]	[CAPRS]	× [CAPRP] +	
14. *Kurti*	[KURTI]	[KURTS]	× [KURTP] +	

Table A.13. Additional inputs for agricultural production.

	Total input	Price (Rs/unit)	Expenditure
Irrigation: (h)			
8. Tubewell		+	[TWCOS]
9. Diesel	[DIESL]×	[DIESP] +	

3.4. Agriculture

All questions are the same as in Section 1.4.. The amount of land is also measured in acres. However, the agricultural output, as mentioned in Table A.1, is different. A few alternative crops are cultivated in Jharkhand. These are known by their local names: *mahua*, *chapra*, and *kurti*. Table A.1 can be extended by including the three agricultural crops as presented in Table A.12.

The main source of irrigation is by natural flows from rivers. Some people reported to have access to a tubewell or a pump. Table A.2 can be extended by including the two agricultural inputs as presented in Table A.13.

3.4.1. Configuration of the Pooler

Are you a pooler of land? (yes/no)
How many acres of land have you pooled?
Table A.14 is used to estimate the expected return from pooling.

3.5. Livestock Rearing

Nobody reported to own any mules or camels, but a few families owned pigs. This is the only difference in cattle stock with Haryana as mentioned in Table A.3.

3.6. Fodder, Fuelwood, Timber and Loans

Free grazing is practised in Jharkhand in an organised form, where the people have to pay a so called grazing fee, so that their cattle can graze freely on the private lands in the slack season.

The following additional questions were asked:

Table A.14. Overview of planted trees in pooled land.

Name of the planted saplings	Initially planted	Finally survived	Value (Rs)
1. Melon	[MELOP]	[MELON] ×	[MELOV] +
2. Papaya	[PAPAP]	[PAPAY] ×	[PAPAV] +
3. Banana		×	+
4. Guava	[GUAVP]	[GUAVA] ×	[GUAVV] +
5. Lemon	[LEMOP]	[LEMON] ×	[LEMOV] +
6. Orange	[ORANP]	[ORAN] ×	[ORANV] +
7. Bamboo	[BAMBP]	[BAMB] ×	[BAMBV] +
8. *Gamhar*	[GAMHP]	[GAMH] ×	[GAMHV] +
9. Sisam	[SISAP]	[SISAM] ×	[SISAV] +
10. Subabul	[SUBAP]	[SUBAB] ×	[SUBAV] +
11. Eucalyptus	[LIPTP]	[LIPTU] ×	[LIPTV] +
12. *Khair*	[KHAIP]	[KHAIR] ×	[KHAIV] +
13. Acacia	[ACACP]	[ACAC] ×	[ACACV] +
14. Teak	[TEAKP]	[TEAK] ×	[TEAKV] +
15. Cashew	[KAJUP]	[KAJU] ×	[KAJUV] +
16. *Margosa* or neem tree	[NEEMP]	[NEEM] ×	[NEEMV] +
17. Pomegranate	[POMEP]	[POMEG] ×	[POMEV] +
18. Mulberry	[MULBP]	[MULB] ×	[MULBV] +
19. Mango	[MANGP]	[MANGO] ×	[MANGV] +
20. *Aunla* (myrobalan)	[AULAP]	[AULA] ×	[AULAV] +
21. Jackfruit	[KATAP]	[KATAL] ×	[KATAV] +
22. Jujube plum	[BERHP]	[BERH] ×	[BERHV] +
23. *Mahua*	[MARUP]	[MARUA] ×	[MARUV] =
Grand total	[PLANT]	[STAND]	[PLVAL]

FREEG = Are you sending your animals for grazing on someone else's private lands? (yes/no)

In Jharkhand, goods are also collected from the pool, while forests are not always there. It is necessary to substitute Table A.6 completely Table A.15.

Loans are either taken from the bank or the shopkeeper. Nobody reported to have taken loans from landlords. As a third possibility, people are also taking loans from the village fund.

3.7. The Configuration of the Family

The collected information about each family member is exactly the same as the ones presented in Section 1.7. and Table A.8.

Table A.15. Sources of fodder, fuelwood, *bhabbar* and bamboo.

(in quintals per year) From:	Fuelwood	Fodder Grass	Fodder Oil-cake
(Reserved) forest	[FWFOR]	[FODFOR]	
Panchayat land	[FWPAN]		
Pool of land	[FWPOO]	[FODPOO]	
Private land	[FWPRV]	[FODPRV]	
Market	[FWMAR]	[FODMAR]	[KHAL]
Total use	[FWTOT]	[FODTOT]	
Market price	[FWPR]	[FODPR]	[KHALP]
Grazing cost		[GRFEE]	

Table A.16. Additional irrigation facilities in Jharkhand.

Capital-goods:	Number/amount	Value	Total value	Sub-total
3. Irrigation:				
a. wells	[WELL]	× [WELLV] +		
b. pumps	[PUMP]	× [PUMPV] +		
c. persian wheels	[RAHAT]	× [RAHAV] +		
d. ponds	[POND]	× [PONDV]=		[CIRRI]

3.8. Consumption

With only a minor alteration, Table A.9 can also be used for Uttaranchal. Nobody reported to own a phone, a gas cylinder, nor had a servant in the household.

3.9. Capital

People from Jharkhand own lesser capital than the people of Haryana, but they own a number of water sources, which they use for small scale irrigation. Nobody reported to own any fridge, cooler, solar, phone, car, truck, electric or diesel fodder cutter, thresher, or a tractor.

The irrigation facilities as shown in Table A.16 should be added to Table A.10.

3.10. Definitions of the Derived Variables

The definitions of Section 1.10. need to be modified for the level of participation, animal income, and the payoff. Two new definitions need to be added: forest accessibility and the survival rate of the plantation in the pool.

FORAV = Forest accessibility: [FUSE] + [FUFAR] + [FGRAZ].

SRATE = Survival rate of the plantation in the pool for the land-pooler: $\frac{[\text{STAND}]}{[\text{PLANT}]}$.

PARTI = Level of participation based on the total sum: [PEXT] + [PBEN] + [PUSE] + [PUBEN] + [GMEET] + [GSINP] + [GPRES] + [GSFOL] + [GSINF] + [GSINT] + [GSUSE] + [GSSUG].

ANICO = Cost of animal husbandry : [FODPR] × [FODFM] + [KHAL] × [KHALP] + [GRFEE].

TCVP = Total salary earned through *chakriya vikas pranali*. (Rs per year)

LANSH = Expected share for the land-pooler (formula A.9).

STUSH = Expected share for the student (formula A.10).

The share of the land-poolers and the students are based on the reported plantation value by the landpoolers. Their expected share is based on the sharing formula in *chakriya vikas*, where the land poolers get 30% of the marketed surplus, and all the students get 30%. N_j is as before the set of interviewed people in village j.[4]

$$[\text{LANSH}] = \begin{cases} \frac{[\text{PLVAL}]}{[\text{PLANY}]} \times 0.3 & \text{if } [\text{PLAND}] > 0 \\ & \text{and } [\text{PWN}] = 1 \\ 0 & \text{if } [\text{PLAND}] = 0 \\ & \text{or } [\text{PWN}] = 2 \end{cases} \tag{A.9}$$

$$[\text{STUD}] = \begin{cases} 1 & \text{if } [\text{PJOB}] \geq 100 \\ 0 & \text{if } [\text{PJOB}] < 100 \end{cases}$$

$$[\text{NRSTUD}]_i = \sum_{i \in N_j} [\text{STUD}]_i$$

$$[\text{STUSH}] = \begin{cases} \frac{\Sigma_{k \in N_j} [\text{LANSH}]_k}{[\text{NRSTUD}]_i} \; i \in N_j & \text{if } [\text{PJOB}] \geq 100 \\ 0 & \text{if } [\text{PJOB}] < 100 \end{cases} \tag{A.10}$$

PAYOF = Total payoff: (consists of three parts) [LANSH] + [STUSH] + [TCVP].

PWN = Pooling indicator . . . (1. pooler/ 2. withdrawer/ 0. non-pooler)

PLYMAX = For how many years has the pooling process been going on in your village?

3.11. Definition of Pooling Intensity

Pooling intensity is defined as the average pooling rate of labour and/or land over the total relevant pooling period.

$$[\text{PISEL}] = \begin{cases} \frac{\frac{[\text{PLAND}]}{[\text{LANDO}]}[\text{PLANY}] + \frac{[\text{PLEXP}]}{[\text{LANDO}]}}{2[\text{PLYMAX}]+1} + \frac{[\text{PJOB}]}{365}\frac{[\text{PJOBY}]}{2[\text{PLYMAX}]+1} & \text{if } [\text{LANDO}] > 0 \\ \frac{[\text{PJOB}]}{365}\frac{[\text{PJOBY}]}{[\text{PLYMAX}]} & \text{if } [\text{LANDO}] = 0 \end{cases}$$

[4]The set of interviewed villagers N_j is not per competing village groups, but per actual village. There are 17 villages in the Jharkhand survey, while the game among villagers is played in 12 different village groups.

4. Village Level Questionnaire for Jharkhand

Who has which function in the *gram samaj*?

1. President (Vice-president);
2. Secretary (Vice-secretary);
3. Casher (Vice-casher);
4. Managing members: they can be member of one of the following committees:
    - Operator student fund;
    - Operator village fund;
    - Executive committee;
    - Social account committee;
    - Protection committee;
    - Marketing committee;

- List all people who worked as a student in *chakriya vikas pranali*.
- Did they work full-time or temporary?
- In which years did they work?

- List all people who pooled land.
- How much land have they contributed?
- When did they pool land?

- What was the date of the last election? (committee-wise)
- What is the total size of the pool under control of the managing committee? (acres)
- What is the total size of irrigated land in the village? (acres)
- What is the over-all quality of the pool? (scaling 1–7)
  Very degraded 1 ...... 2 ...... 3 ...... 4 ...... 5 ...... 6 ...... 7 very good
- How far is the forest to collect primary resources from the village? (km)
- How many communities/villages depend upon that forest? (number)
- Is there any coordination between these dependent villages? (yes/no)
- Is free grazing happening in the forest? (yes/no)
- Is free grazing happening in the pool? (yes/no)
- Has the situation of free grazing changed? (Improved/no/worsened)
- What is done against free-grazing?
- What is done against people who let their animals graze freely?
- Is there any problem of encroachment in the reserved forest? (yes/no)
- What are the main problems in the management of the pool?

The general village level indicators like demographics, land- and cattle-ownership and capital, are exactly the same as displayed in Table A.11, only the caste composition differs.

- How many trees have been planted in the pool at the village level?
- How many saplings have survived till date?
- What is the state of the trees in the pool? ... (good/ medium/ poor)

Table A.17. Survival and condition of trees planted in the common pool.

Name of tree	Planted	Survived	Condition (good/medium/bad)	fruit yield
Papaya				
Guava				
Lemon				
Cashew				
Sesam				
Aunla				
Bamboo				
Pomegranate				
Teak				
Eucalyptus				
Subabul				
Mango				
Khair				
Neem				

- To which extend have the fruit trees borne fruit? (see Table A.17)
- How are the benefits distributed among all individuals?
- How does the village level sharing mechanism affect the individual?
- Which saplings have already led to a marketable surplus?
- Who benefitted by the sales of these benefits?
- How are the benefits from the first tier: i.e. vegetables, grass, and grains distributed among the individuals?
- How much has been marketed?
- How much has been distributed in quantities among the students?

5. Household level Questionnaire for Uttaranchal

The quantitative questionnaire for the *van panchayat*s in Uttaranchal is structurally the same as the questionnaire for Haryana. The basic differences are that the people in Uttaranchal harvest different crops, they obtain fuelwood, fodder, and timber from different types of forests, and their agricultural process is not yet mechanised.

5.1. Basic Characteristics to Identify the Household

The set of questions used here are exactly the same as in Section 1.1..

5.2. Attitude Towards the Environment

All questions are the same as in Section 1.2., except for the position in the local managing body.

VPMEM = If you are a member of the *van panchayat*, then what is your position? ... (0. None. 1. Member, 2. Managing member, 3. Guard, 4. *Pradhan* (*gram Sabha*), 5. *Sarpanch* (*van panchayat*), 6. Secretary (*mahila mangal dal*), 7. Treasurer (*mahila mangal dal*))

5.3. Indicators of Participation

In this Section the participatory responses needed to be reformulated with respect to the local managing body in Uttaranchal: the *van panchayat*.

The following response-pattern is used again:

Not at all	little bit	medium	quite a lot	very much
1	2	3	4	5

Contributing to forest conservation:

FPLANT = To which extent do you plant trees in the reserved forest?

VPLANT = To which extend do you plant trees in the communal forest?

VPCON = Are you contributing something like money or voluntary labour to the *van panchayat* to help in its functioning?

Sharing in the benefits:

VPBEN = To which extent are you benefitting from the presence of the *van panchayat*?

Involvement in decision-making and evaluation through meetings:

VMEET = How frequent are the meetings of your *van panchayat*?

VMIMP = How important do you consider the meetings for proper functioning of your *van panchayat*?

VPRES = How frequent do you attend meetings of the *van panchayat*?

VPFOL = To which extent can you agree with decisions which were made in meetings of the *van panchayat*?

VPINFL = To which extent can you influence the decisions in meetings of the *van panchayat*?

VPINT = To which extent do you find the discussions in meetings of interest?

VPUSE = To which extent do you gain from such meetings?

VPSUG = To which extent do you make suggestions during such meetings?

OWNER = Who is the owner of the forest? ... (1. Government, 2. *Sarpanch*, 3. Villagers, 4. None of them (other).)

Table A.18. Additional outputs from agricultural production.

Total (in quintals)		Amount sold		Price (Rs)		Total yield
12. Soya bean	[SOYA]	[SOYAS]	×	[SOYAP]	+	
13. *Mandwa*	[MANDW]		×		+	
14. *Sanwa*	[SANWA]	[SANWS]	×	[SANWP]	+	
15. *Jav*	[JAV]		×		+	

5.4. Agriculture

All questions are the same as in Section 1.4.. The main alteration is that the amount of land is measured in the local units: *nalis*[5]. Furthermore, the agricultural output, as mentioned in Table A.1, is slightly different. A few alternative grains are cultivated in Uttaranchal, which are more drought resistant, but of a lesser quality, these are know by the local names: *mandwa* (=*koda*), *Sanwa* (=*jhangora*), and *jav*. Soya beans are also cultivated. Table A.1 can be extended by allowing for four more agricultural crops (see Table A.18).

Finally, the main source of irrigation is by natural flows from rivers. Nobody reported to have access to dams or tubewells.

5.5. Livestock Rearing

Nobody reported to own any mules or camels, but a few families owned horses. This is the only difference with the cattle stock of Haryana as it is mentioned in Table A.3.

5.6. Fodder, Fuelwood, Timber and Loans

While free grazing was more or less absent in Haryana, it still prevails in Uttaranchal. The following additional questions were asked:

GRAZE = Is your cattle sent for grazing? (yes/no)
GRAPL = Where is your cattle sent for grazing?
GRMON = For how many months do you send your cattle for grazing?

Fodder which enhances the milk production, like feed and oil cake is hardly used by the people surveyed in Uttaranchal. This question was, therefore, omitted.

In Uttaranchal are found three types of forest from which the people collect three major resources: fodder, fuelwood and timber. It is necessary to substitute Table A.6 completely by Table A.19.

Loans are either taken from the bank or the shopkeeper. Nobody reported to have taken loans from landlords.

5.7. The Configuration of the Family

The collected information about each family member is exactly the same as the ones presented in Section 1.7. and Table A.8.

[5]One ha = 50 *nalis*

Table A.19. Sources of fodder, fuelwood, *bhabbar* and bamboo.

From: (in quintals per year)	Fodder	Fuelwood	Timber
Reserved forest	[FODFOR]	[FWFOR]	[TIMFOR]
Communal forest	[FODVP]	[FWVP]	[TIMVP]
Civil/soyam forest	[FODCIV]	[FWCIV]	[TIMCIV]
Private land	[FODPRV]	[FWPRV]	[TIMPRV]
Market	[FODMAR]	[FWMAR]	[TIMMAR]
Total use	[FODTOT]	[FWTOT]	[TIMTOT]
Market price	[FODPR]	[FWPR]	[TIMPR]

5.8. Consumption

With only a minor alteration, Table A.9 is also used for Uttaranchal. Nobody reported to own a phone, nor had a servant in the household.

5.9. Capital

The people of Uttaranchal reportedly own lesser capital than the people of Haryana. Table A.10 can be shortened and replaced by Table A.20.

5.10. Definitions of the Derived Variables

All the definitions of Section 1.10. hold here as well. Only the level of participation and the payoff are different.

PARTI = Level of participation based on the total sum: [VPLANT] + [VPCON] + [VPBEN] + [VMEET] + [VMINP] + [VPRES] + [VPFOW] + [VPINFL] + [VPINT] + [VPUSE] + [VPSUG].

PAYx = Total payoff: 113.16 × [FODx] + 51.73 × [FWx] + 12.57 × [TIMx]; for all x ∈ {FOR, VP, CIV, TOT}.

5.11. Definition of Forest Dependence

Forest dependence is defined as the amount of goods collected from all types of forest divided by the total need.

$$
\begin{aligned}
&\text{Let } x \in \{\text{FW, FOD, TIM}\} \\
&[\text{xDEP}] = \begin{cases} \dfrac{[\text{xFOR}] + [\text{xVP}] + [\text{xCIV}]}{[\text{xTOT}]} & \text{if xTOT} > 0 \\ 0 & \text{if xTOT} = 0 \end{cases} \\
&[\text{xNR}] = \begin{cases} 1 & \text{if xTOT} > 0 \\ 0 & \text{if xTOT} = 0 \end{cases} \\
&[\text{FORDEP}] = \frac{[\text{FWDEP}] + [\text{FODDEP}] + [\text{TIMDEP}]}{[\text{FWNR}] + [\text{FODNR}] + [\text{TIMNR}]}
\end{aligned}
\qquad (A.11)
$$

Table A.20. Amount and value of privately owned capital goods.

	Capital-goods:	Number	Value	Total value	Sub-total
1.	House			+ [HOUSE]	
	Shop goods			+ [SHOPG]	
a.	*charpai*	[CHARP] ×	[CHARV] +		
b.	bed	[BED] ×	[BEDV] +		
c.	chair	[CHAIR] ×	[CHAIV] +		
d.	bench	[BENCH] ×	[BENCV] +		
e.	table	[TABLE] ×	[TABLV] +		
f.	television	[TV] ×	[TVV] +		
g.	radio	[RADIO] ×	[RADIV] +		
h.	tape-recorder	[TAPE] ×	[TAPEV] +		
i.	cycle	[CYCLE] ×	[CYCLV] +		
j.	scooter	[SCOOT] ×	[SCOOV] +		
k.	motor	[MOTOR] ×	[MOTOV] +		
l.	electric fitting	[ELECF] ×	[ELECV] +		
m.	gas-cylinder	[GASC] ×	[GASV] +		
n.	solar	[SOLAR] ×	[SOLAV] +		
o.	sewing machine	[SEWIN] ×	[SEWIV]=		[CHOUS]
2.	Land and cattle:				
a.	land	[LANDO] ×	[LANDV] +		
b.	oxen	[OX] ×	[OXV] +		
c.	cows	[COW] ×	[COWV] +		
d.	buffaloes	[BUFF] ×	[BUFFV] +		
e.	goats	[GOAT] ×	[GOATV] +		
f.	chicken	[CHICK] ×	[CHICV] +		
g.	horses	[HORSE] ×	[HORSV]=		[CANLA]
4.	Agricultural implements:				
a.	plough and leveller			+ [PLLEV]	
b.	sickle and axe			+ [AXEGD]=	[CGOOD]

6. Village Level Questionnaire for Uttaranchal

Who has which function in the *van panchayat*?

1. president: [SARPANCH]
2. managing members: [MM1], [MM2], [MM3], [MM4], [MM5], [MM6], [MM7], [MM8], [MM9];
3. guards: [GUARD1], [GUARD2];

The general village level indicators like demographics, land- and cattle-ownership and capital, are exactly the same as displayed in Table A.11, only the caste composition differs.

SAL = What is the salary of the guard? (Rs per month)
VILFUND = What the amount in the village account? (Rs)
ELDATE = What is the date of the last election in the *van panchayat*?
REGDATE = What is the registration date of the *van panchayat*?
PRADAN = What is the name of the *pradhan*.
NRMMGS = What is the number of managing members in the *gram sabha*?

NRFMGS = What is the number of female members in the *gram sabha*?

PRES_MMD = What is the name of the president of the *mahila mangal dal*?

NRMMD = What is the number of managing members in the *mahila mangal dal*?

VPLAND = What is the size of the communal forest under the control of your *van panchayat*? (hectare)

CSLAND = What is the size of the civil forest? (hectare)

RFLAND = What is the size of the reserved forest which the people of your village are using? (hectare)

PRLAND = What is the total size of private land in your village? (hectare)

FAMILY = What is the number of families in your village?

POPUL = What is the population of your village?

OX = How many oxen are there in your village?

BUFF = How many buffaloes are there in your village?

COW = How many cows are there in your village?

GOAT = How many goats are there in your village?

HORSE = How many horses are there in your village?

WELL = How many wells are there in your village?

GASCIL = How many gas cylinders are there in your village?

SCHOOL = Up to which class are schools available in your village?

Appendix B

Collection of Data Tables

1. Organisations in Haryana & Jharkhand

Table B.1. Overview of registered Hill Resource Management Societies until 16-02-90.

registration				house	popu-	forest	fod-	bha-	wa-	bam-	dams	
date	no.	Village name	Range	holds	lation	(ha)	der	bhar	ter	boo	#	ok
01-07-83	251	Lohgarh	Pinjore	70	540	231.6	T	T	T	F	2	2
01-07-83	252	Dhamala	Pinjore	239	1565	161.1	T	T	T	F	2	2
01-07-83	254	Sukhomajri	Pinjore	100	700	151.6	T	T	T	F	4	1
01-07-83	253	Jattan Majri	Pinjore	66	555	131.6	T	T	F	F	2	0
26-08-83	401	Kheda	Pinjore	82	500	126.0	T	T	F	F	1	
11-12-84	544	Chowki	Panchkula	0	0	374.2	F	F	T	F	2	1
05-08-85	292	Harijan Nada Sahib	Panchkula	20	125	196.3	T	F	F	F	2	0
01-12-85	653	Nada Sahib	Panchkula	279	1980	682.3	F	F	T	F	6	4
29-05-87	833	Masoompur	Raipur Rani	41	305	281.1	T	T	F	F	2	0
29-05-87	832	Raina	Raipur Rani	130	882	370.8	T	T	F	F	0	0
29-05-87	835	Dulopur	Raipur Rani	12	128	495.7	F	F	F	F	3	
29-05-87	834	Tirlokpur	Raipur Rani	68	411	751.9	T	F	F	F	1	
27-06-89	236	Surajpur	Pinjore	464	2551	189.5	T	T	T	F	1	1
07-07-89	309	Uttamwala	Chachrauli	22	143	212.0	T	F	T	F	1	
07-07-89	306	Faizpur	Kalesar	150	950	466.2	T	F	T	F	2	
07-07-89	310	Gobindpur/ Mandpa	Raipur Rani	72	537	363.0	T	T	T	F	1	1
07-07-89	308	Paniwala	Chachrauli	25	171	60.0	T	F	T	F	3	3
07-07-89	307	Rampur Gainda	Sadhaura	15	100	85.0	T	T	T	F	1	
05-11-89	886	Mirpur	Raipur Rani	67	458	547.6	T	T	F	F	0	0
20-11-89	966	Kahinwala	Pinjore	62	452	412.9	T	T	F	F	0	0
02-01-90	1063	Bagpat	Kalesar	270	1754	529.3	T	T	T	F	2	2
09-01-90	1082	Khokhra-Thappal	Pinjore	85	578	257.8	F	F	T	F	2	
15-01-90	1099	Khillanwala	Kalesar	180	1034	867.6	T	T	T	F	2	
02-02-90	1158	Godam	Pinjore	61	354	196.0	F	F	F	T	0	0
16-02-90	1219	Thaska	Naraingarh	70	698	338.4	T	F	T	F	1	1

Note: T = true and F = false.

Sources: Data are obtained from TERI, cross-checked with: [94] and [210].

This Section presents a wide array of characteristics of villages where Hill Resource Management Societies are active in Haryana and villages under *chakriya vikas prenali* in Jharkhand. Table B.1 and Table B.2 present the created Hill Resource Management Organisations in Haryana, while Table B.3 and Table B.4 give an overview of villages under *chakriya vikas prenali*, until 1996. A detailed overview of the 4500 registered *van panchayats* is not attempted here. Additional statistics on *van panchayats* can be found

Table B.2. Overview of registered Hill Resource Management Societies from 17-02-90 onwards.

registration date	no.	Village name	Range	house holds	popu-lation	forest (ha)	fod-der	bha-bhar	wa-ter	bam-boo	dams #	ok
26-02-90	1239	Kalka	Pinjore	36	252	329.3	F	F	F	T	0	0
26-02-90	1240	Gumthala	Panchkula	38	208	301.5	F	F	F	F	2	0
12-03-90	1294	Mandhana	Morni	330	2409	0.0	F	F	T	F	1	0
02-06-90	310	Khizri	Kalesar	200	1326	256.0	T	T	T	F	2	
29-07-90	331	Salehpur	Sadhaura	106	749	362.0	T	F	T	F	2	1
06-08-90	483	Pammuwala	Sadhaura	75	521	89.0	T	F	T	F	2	
19-02-91	1044	Thadion	Raipur Rani	72	526	658.8	T	F	T	F	1	
19-02-91	1043	Bharauli	Raipur Rani	234	1658	712.3	F	F	T	F	1	
25-10-91	665	Mauliwal-Bhud	Raipur Rani	136	924	812.2	T	T	T	F	1	
07-07-92	518	Thathar/ Madhowala	Raipur Rani	72	526	353.7	T	T	F	F	0	0
23-02-93	1702	Manakpur Thakurdas	Pinjore	82	517	178.2	T	T	F	F	0	0
09-04-93	38	Khuda Baksh	Pinjore	46	331	108.4	F	F	F	F	0	0
20-06-94	389	Manakpur Nanak Chand	Pinjore	102	702	229.1	T	T	F	F	0	0
31-08-94	654	Rajipur	Pinjore	714	3844	235.9	T	T	F	F	0	0
31-08-94	666	Nanakpur	Pinjore	12	0	124.9	T	T	F	F	1	
01-09-94	771	Kiratpur	Pinjore	249	1485	238.8	T	F	T	F	1	
02-02-95	1479	Basaulan	Pinjore	157	1036	107.3	T	T	F	F	1	
18-05-95	292	Kathgarh	Sadhaura	140	879	215.7	T	T	T	F	3	
18-05-95	291	Bhagwanpur	Sadhaura	120	771	2.0	T	T	T	F	3	
24-07-95	774	Prem Pura	Pinjore	42	274	153.3	T	T	F	F	1	1
	0	Kona	Pinjore	0	0	200.0	T	F	F	F	0	0
	0	Kedarpur	Pinjore	0	0	0.0	F	F	F	F	0	0
	0	Damdama	Pinjore	0	0	0.0	F	F	F	F	2	
Total:				5613	36409	14148	36	25	25	2	67	20
Per village:				117	759	295	$\frac{3}{4}$	$\frac{25}{48}$	$\frac{25}{48}$	$\frac{1}{24}$	2.7	

Note: T = true and F = false.

Sources: Data are obtained from TERI, cross-checked with: [94] and [210].

in Pandey [157] and Saxena [178]

2. Inter State Comparative Data

This Section presents the aggregated average household level data of the primary data survey and it is linked with village level data. First, Table B.5 presents the inter-State variation among gender, age, education, and employment. Second, Table B.6 presents the inter-State variation in caste, management position, and the perceived reason for forest degradation. Third, Table B.7 presents an inter-State comparison of demographics and inequalities. Fourth, Table B.8 presents an inter-State comparison of income, consumption and capital.

3. Intra State Comparative Data

This Section presents the aggregated average household level data as gathered during the field survey at the village level. First, Tables B.9–B.12 present information about the situation in Haryana. Second, Tables B.13–B.16 present information about the situation in Jharkhand. Third, Tables B.17–B.20 present information about the situation in Uttaranchal.

The same numbering as in Section 2. is used here to make a link between the data at the State and village level. More specifically the following information is presented in the Tables:

Table B.3. Overview of active villages under cyclic development in Daltenganj and Giridih.

No	Village name	years	plantation activities: area (acre)	seed-lings	sa-plings	grafts	water harvesting: tie-ridges	tank	dam	well	village & student fund (Rs)
Daltenganj sub-division:											
1	Bhusaria	87-94	135	280350	187031	17502	27199	7	23	3	76180
2	Tandwa	88-94	105	300500	229752	4979	14523	4	0	0	29295
3	Jorkat	88-94	132	306750	235665	3700	19565	5	3	1	51955
4	Kauria	88-94	123	244700	180459	600	19980	1	2	2	37665
5	Awsane	88-94	91	220500	130861	1050	32634	0	5	2	27283
6	Kechki	88-94	128	384500	255366	7124	29799	0	4	0	66779
7	Kachanpur	88-94	108	257500	191323	1800	31503	2	3	1	40640
8	Pirhia	89-94	101	208800	165394	1700	9905	7	0	0	34311
9	Pokhraha	89-94	102	272100	176302	4350	9885	3	3	2	34527
10	Sakanpirhi	88-94	103	263700	155145	1500	16450	2	0	0	24926
11	Chukru	92-94	22	37300	38040	0	2500	0	0	0	12868
12	Kashia	94-96	45	39000	60050	0	3100	0	0	0	10829
13	Khamdih	94-96	41	42000	41037	0	1450	0	0	0	9392
14	Bakhari	94-96	41	45000	60890	0	1772	0	0	0	10208
15	Mundaria	94-96	41	35000	57210	0	3500	0	0	0	10626
16	Kumbhawa	94-96	40	60000	70000	0	2800	0	3	0	11441
Giridih (Aurangabad district in Bihar):											
39	Bagda	88-90	23	49000	39244	0	0	1	0	0	1492
40	Bagiari	88-90	96	235000	180988	0	0	4	0	0	5598
41	Mungo	88-90	25	65000	42167	0	0	1	0	0	2476
42	Tangtona	88-90	34	90000	59369	0	0	2	0	0	2885
43	Jamkudar	88-90	9	42500	14215	0	0	1	0	0	1277
44	Belsara	94-96	1	50000	40500	0	2820	0	0	0	4145

Sources: SHRMS [186], [187], [188], [189], and Sinha [194].

Table B.4. Overview of active villages under cyclic development in Barwadih.

No	Village name	years	plantation activities: area (acre)	seed-lings	sa-plings	grafts	water harvesting: tie-ridges	tank	dam	well	village & student fund (Rs)
Barwadih sub-division:											
17	Chapri/ Karmahi	87-94	169	331700	239231	29243	31525	5	18	0	195709
18	Khura	87-94	166	283000	245932	2590	14277	2	11	0	67954
19	Lanka	87-94	129	251550	200598	2629	14572	1	5	0	59703
20	Sindhorwa	88-94	95	209100	163426	1155	17766	2	0	1	19400
21	Putuagarh	88-94	78	175250	160461	50	27512	6	1	0	18226
22	Chhecha	89-94	78	173450	141208	1795	6000	3	1	0	20913
23	Muru	88-94	109	208400	168755	1092	22055	3	3	0	22839
24	Dorami	88-94	110	227650	165781	1671	22383	2	0	1	23462
25	Mangra	88-94	120	221200	168992	0	23775	4	1	1	22094
26	Kurka	89-94	200	453100	385873	4600	94161	6	0	0	49674
27	Chapri School	92-94	19	20500	28118	0	2176	0	0	1	8585
28	Chhecha School	88-94	81	186300	94545	5886	16750	1	0	0	24482
29	Horilong School	88-90	31	91600	40763	0	4500	5	0	0	12338
30	Marhi Chhech-hani	90-91	10	0	9519	0	0	0	0	0	0
31	Morwai	94-96	4	25000	50834	0	3500	0	0	0	8991
32	Hondohao	94-96	18	15000	59725	0	7000	0	0	0	9551
33	Baheratanr	94-96	4	23000	63639	0	3500	0	0	0	9494
34	Murgidih	94-96	8	45000	36280	0	7000	0	0	0	9162
35	Karkatia	94-96	43	21000	59848	0	5250	0	0	0	9841
36	Barhania	94-96	88	70000	80000	0	10000	3	0	0	12732
37	Khura Lanka	95-96	1	50000	0	0	0	1	0	0	839
38	Chhechha-Horilong	95-96	1	40000	0	0	0	1	0	0	2200
	Total		3137	6681000	5174536	95016	563087	85	86	15	1114991
	Per village		71	151841	117603	2159	12797	2	2	0	25341

Sources: SHRMS [186], [187], [188], [189], and Sinha [194].

Table B.5. Inter State comparison of gender, age, education, and employment.

State	Haryana		Jharkhand		Uttaranchal	
Interviewed persons:	freq	%	freq	%	freq	%
Gender: Male	118	92.9	113	91.9	123	91.1
Female	9	7.1	10	8.1	12	8.9
Age groups: 11-20	13	10.2	10	8.1	7	5.2
21-30	41	32.3	36	29.3	17	12.6
31-40	34	26.8	36	29.3	30	22.2
41-50	22	17.3	22	17.9	36	26.7
51-60	10	7.9	10	8.1	22	16.3
61-100	7	5.5	9	7.3	23	17.0
Education groups:						
0 Illiterate	48	37.8	34	27.6	20	14.8
1–4 Literate	6	4.7	24	19.5	9	6.7
5–6 Primary pass	16	12.6	7	5.7	25	18.5
7–9 Middle pass	26	20.5	32	26.0	30	22.2
10–11 Matric pass	24	18.9	16	13.0	17	12.6
12–14 Inter College pass	4	3.1	7	5.7	17	12.6
15–20 Graduated	3	2.4	3	2.4	17	12.6
Families of interviewed persons:						
Gender: Male	517	55.7	480	49.1	436	51.1
Female	411	44.3	497	50.9	417	48.9
Age groups: 0-10	255	27.5	326	33.4	217	25.4
11-20	201	21.7	206	21.1	211	24.7
21-30	195	21.0	174	17.8	137	16.1
31-40	114	12.3	100	10.2	89	10.4
41-50	72	7.8	58	5.9	79	9.3
51-60	47	5.1	59	6.0	55	6.4
61-100	44	4.7	54	5.5	65	7.6
Education groups:						
0 Illiterate	439	47.3	473	48.4	256	30.0
1–4 Literate	143	15.4	213	21.8	136	15.9
5–6 Primary pass	173	18.6	110	11.3	155	18.2
7–9 Middle pass	72	7.8	80	8.2	116	13.6
10–11 Matric pass	82	8.8	67	6.9	95	11.1
12–14 Inter College pass	13	1.4	25	2.6	61	7.2
15–20 Graduated	6	0.6	9	0.9	34	4.0
Employment:						
Unemployed	22	2.4	3	0.3	53	6.2
Expected or going to school	355	38.3	412	42.2	395	46.3
Household activities	198	21.3	281	28.8	63	7.4
Primary sector	206	22.2	243	24.9	250	29.3
Industrial sector	90	9.7	9	0.9	8	0.9
Service sector	57	6.1	29	3.0	84	9.8

Table B.6. Inter State comparison of caste, management position, and the reason for forest degradation.

State	Haryana		Jharkhand		Uttaranchal	
Caste distribution:	freq	%	freq	%	freq	%
Brahmin: Pandit	2	1.6	19	15.4	10	7.4
Ksatriya: Rajput	12	9.4	3	2.4	114	84.4
Jat	10	7.9	0	0.0	0	0.0
Baiyas: Gujjar	55	43.3	0	0.0	0	0.0
Sudra: Lohar	1	0.8	3	2.4	3	2.2
Other	0	0.0	13	10.6	1	0.7
Scheduled castes(SC):	2	1.6	7	5.7	4	3.0
Bhanjda	30	23.6	0	0.0	0	0.0
Harijan	11	8.7	0	0.0	2	1.5
Scheduled tribes(ST):						
Munda	0	0.0	22	17.9	0	0.0
Bunia	1	0.8	14	11.4	0	0.0
Chero	0	0.0	14	11.4	0	0.0
Parahia	0	0.0	10	8.1	0	0.0
Korva	0	0.0	7	5.7	0	0.0
Uraun	0	0.0	6	4.9	0	0.0
Other	2	1.6	3	2.4	0	0.0
Neither caste nor tribal	1	0.8	2	1.6	1	0.7
Position in managing body:						
No member	3	2.4	1	0.8	3	2.2
Common member	68	53.5	50	40.7	89	65.9
Managing member	36	28.3	41	33.3	16	11.9
Guard	0	0.0	0	0.0	3	2.2
Casher	7	5.5	8	6.5	1	0.7
Secretary	4	3.1	8	6.5	2	1.5
President	9	7.1	15	12.2	21	15.6
Reason for forest degradation:						
Negligence by people	34	26.8			37	27.5
Over-population	4	3.1			35	26.0
Destruction by people	2	1.6			17	12.6
Fires	17	13.4			11	8.1
Conflicts in community	8	6.3			6	4.4
Negligence by government	12	9.4			3	2.2
Natural: bad monsoon	16	12.6			0	0.0
Wild animals	3	2.4			0	0.0
Other reasons	11	8.7			11	8.1
Not degraded at all	20	15.7			15	11.1

Table B.7. Inter State comparison of demographics and inequalities.

State	Haryana	Jharkhand	Uttaranchal
1. Demographics: (averages at the village level)			
Number of households	79	70	148
Averages in the field survey:			
Number of households	14	10	12
Number of family members	103	81	78
Number of women	46	41	38
Number of adults	59	44	44
Number of children	44	37	34
Female-male ratio	886	1017	992
Number of people employed	62	47	54
employed outside the village	12	6	9
Years of schooling (age above 15)	3.47	3.52	6.09
2. Inequalities:			
2a. Caste:			
Brahmin (highest caste)	2	19	10
Ksatriya (fighters caste)	22	3	114
Baiyas (landowners/traders caste)	55	0	0
Sudra (lowest caste)	48	25	11
Tribals		76	
Equality of caste	0.360	0.699	0.725
2b. Religion:			
Hindu	114	99	134
Muslim	1	2	0
Sikh	12	0	0
Christian	0	22	1
Equality of religion	0.815	0.680	0.985
Equality index	0.588	0.690	0.855
Gini-coefficient	0.217	0.150	0.239

Table B.8. Inter State comparison of income, consumption and capital.

State	Haryana	Jharkhand	Uttaranchal
4. Income:			
4a. Average household level income			
From land	5564	1432	1276
Land income per acre	2092	313	1081
From cattle	2996	39	794
Cattle income per cattle	687	15.4	286
Outside village income	13746	5995	19912
Within village income	14163	1846	7333
through *chakriya vikas prenali*	0	3453	0
Total household income	27909	12766	27245
Per capita income	3954	1622	5200
4b. Average output (quintals per year per household):			
Agriculture	30.2	16.2	16.3
Yield rate per acre	11.3	3.6	13.8
Animal husbandry	27.8	1.4	5.8
Yield rate per cattle	6.4	0.6	2.1
4c. Average payoffs (Rs per year):	6555	10927	2228
4d. Village fund (Rs per year):	15559	36928	16933
5. Consumption			
Fodder per cattle (qu/year)	32.2	4.2	14.6
Fuelwood per household (qu/y)	14.3	33.2	23.9
Fuelwood per capita (qu/y)	2.0	4.3	3.6
Food per household (Rs/year)	21126	10998	20657
Food per capita (Rs/year)	3085	1573	3763
6. Capital: (per household)			
Forest in use (ha)	470	129	730
Size of the pool (acres)	0	56	0
Land-ownership:			
Land owned (acres)	3.19	6.17	1.41
Land cultivated (acres)	2.65	4.57	1.18
Land irrigated (acres)	1.46	1.22	0.18
Number of landowners	0.66	0.93	0.99
Cattle-ownership (number):			
Buffaloes	2.48	0.24	0.75
Cows	0.54	0.92	1.55
Goats	0.80	1.54	1.06
Draught animals	0.97	1.83	1.39
Number of cattle owners	0.72	0.62	0.96
Average number of milch animals	4.36	2.56	2.78
Per capita capital (Rs)	1,47,000	26,000	46,000
Household capital (Rs)	10,53,000	2,02,000	2,38,000

- Table B.9 presents demographics and inequalities.
- Table B.10 presents forest-ownership and income.
- Table B.11 presents consumption and capital.
- Table B.12 presents irrigation facilities and village level indicators.
- Table B.13 presents demographics and inequalities.
- Table B.14 presents a classification of poolers and income.
- Table B.15 presents consumption, land and cattle-ownership.
- Table B.16 presents private assets, irrigation facilities and village level indicators.
- Table B.17 presents demographics, inequalities, forest-ownership and average output.
- Table B.18 presents income and benefits from participation.
- Table B.19 presents consumption and capital.
- Table B.20 presents irrigation facilities and village level indicators.

Table B.9. Demographics, forest-ownership and inequalities in Haryana.

Village group	1	2	3	4	5	9	6	7	8
1. Demographics:									
Number of households	33	43	90	128	162	20	54	52	130
Averages in the field survey:									
Number of households	30	11	19	16	11	4	12	14	10
Number of family members	179	86	114	162	92	25	84	100	86
Number of women	79	36	45	76	43	12	36	39	45
Number of adults	89	60	70	98	52	19	42	55	46
Number of children	90	26	44	64	40	6	42	45	40
Female-male ratio	883	837	789	938	935	960	857	780	1047
Number of people employed	96	60	69	102	57	18	50	59	47
employed outside the village	7	6	23	22	11	7	10	15	8
Years of schooling (age above 15)	2.70	3.45	4.21	4.25	3.64	6.25	3.08	1.86	4.30
2. Inequalities:									
2a. Caste:									
Brahmin (pandit)	0	0	0	2	0	0	0	0	0
Ksatriya: Rajput	0	0	0	0	11	1	0	0	0
Jat	0	0	0	10	0	0	0	0	0
Baiyas: Gujjar	0	11	18	0	0	0	10	8	8
Sudra: Harijan	0	0	0	2	0	3	0	5	1
Bhanjda	30	0	0	0	0	0	0	0	0
Other (low caste)	0	0	1	2	0	0	2	1	1
Equality of caste	1.000	1.000	0.900	0.430	1.000	0.625	0.722	0.510	0.680
2b. Religion:									
Hindu	30	11	19	5	10	3	12	14	10
Sikh	0	0	0	10	1	1	0	0	0
Muslim	0	0	0	1	0	0	0	0	0
Equality of religion	1.000	1.000	1.000	0.492	0.835	0.625	1.000	1.000	1.000
Equality index	1.000	1.000	0.950	0.461	0.918	0.625	0.861	0.755	0.840
Gini-coefficient	0.118	0.225	0.204	0.151	0.235	0.114	0.254	0.202	0.281
3. Owner of the forest:									
Government		6	3	2	7	1	6	9	9
Society		2	9	4	2	3	2	3	1
Government and society	1		4	7	1	0	3	2	0
Other			2	2	1	0		0	0

Note: Village groups: 1=Godam, 2=Mandpa, 3=Sukhomajri, 4=Dhamala, 5=Main Nada, 9=Harijan Nada, 6=Masoompur, 7=Thathar, 8=Salehpur.

Table B.10. Various types of income in Haryana.

Village group	1	2	3	4	5	9	6	7	8
4. Income: (per year)									
4a. Average household level income:									
From land	0	10012	5974	4985	10790	31	6511	-50	20691
Land income per acre	0	910	314	312	981	8	543	-4	2069
From cattle	-249	19320	-249	13019	-1783	-9600	3097	-647	184
Cattle income per cattle	-50	134	-4	133	-32	-1920	40	-10	4
Outside village income	2048	2636	27700	26231	21727	34500	10900	6721	10740
Within village income	16791	32539	8073	19831	9553	-9569	9691	4089	22605
Total household income	18839	35176	35772	46063	31280	24931	20591	10810	33345
Per capita income	3381	4706	6151	4688	4363	4227	2948	1829	3116
4b. Average output: (in quintals per year)									
Agriculture	10	611	615	475	537	60	430	161	928
Yield rate per acre	0.0	9.0	11.8	9.9	10.3	15.0	8.4	9.5	20.6
Animal husbandry	11	900	568	1010	309	32	313	141	257
Yield rate per cattle	3.4	7.4	10.4	11.4	8.2	5.7	5.8	3.4	6.8
4c. Average benefits from participation:									
Damwater (hours)	0.0	58.8	6.7	16.6	13.4	2.5	0.0	0.0	3.4
Fodder from forest (qu)	1.0	28.5	33.5	24.4	27.3	17.5	22.8	28.9	28.5
Bhabbar from forest (kg)	0	11	26	28	82	18	46	18	120
Bamboo from forest (#)	840	0	0	0	0	0	0	0	0
Fuelwood from forest (qu)	10.0	9.7	8.1	6.5	14.1	12.5	11.9	18.4	26.8
Average payoff (Rs)	4162	16048	6253	6960	7684	3987	4150	5486	7387
4d. Benefits at the village level: (Rs)									
Water charges	0	5000	4050	3780	10000	0	0	0	3550
Fodder charges	0	1000	10000	-415	0	0	1000	1000	21000
Bhabbar lease	0	20000	41000	34019	0	0	50000	15000	234000
Fish lease	0	0	2600	3500	10000	0	0	0	0
Total HRMS income	0	26000	57650	40884	20000	0	51000	16000	258550
Village fund	11877	10000	5442	26643	10000	5068	55000	16000	0

Note: Village groups: 1=Godam, 2=Mandpa, 3=Sukhomajri, 4=Dhamala, 5=Main Nada, 9=Harijan Nada, 6=Masoompur, 7=Thathar, 8=Salehpur.

Table B.11. Consumption and capital in Haryana.

Village group	1	2	3	4	5	9	6	7	8
5. Consumption (per year)									
Fodder per cattle (qu)	31.7	17.6	38.3	38.7	40.9	49.0	24.9	18.5	39.8
Fuelwood per household (qu)	11.1	14.7	8.1	11.6	14.1	15.0	13.5	20.3	31.8
Fuelwood per capita (qu)	1.87	1.88	1.35	1.14	1.68	2.40	1.93	2.84	3.70
Food per household (Rs)	16776	20513	19646	27013	27610	20153	12555	31739	27469
Food per capita (Rs)	2957	2698	3470	3116	3761	4529	2980	1866	3625
6. Capital:									
6a. Land-ownership:									
Forest in use (ha)	194	400	114	116	2500	17	120	400	373
Land owned (acres)	0	85	50	62	77	11	52	19	49
Land cultivated (acres)	0	68	52	48	52	4	51	17	45
Land irrigated (acres)	0	30	27	23	37	4	42	2	21
Number of landowners	0	11	18	12	10	4	11	10	8
6b. Cattle-ownership: (number)									
Buffaloes	1	91	44	73	29	3	36	13	25
Cows	2	13	2	1	3	2	11	26	8
Goats	12	63	0	15	0	0	0	12	0
Draught animals	2	22	12	10	18	0	24	25	10
Number of cattle owners	2	11	17	13	11	4	11	12	10
Average number of milch animals per household	1.5	11.1	3.2	6.8	3.4	1.3	4.9	3.4	3.8
6c. Private-assets:									
Ox cart	0	3	0	3	0	0	4	0	4
Plough	1	10	12	8	7	0	11	10	5
Tubewells	0	1	2	2	1	0	1	1	3
Cane crushers	0	1	1	1	0	0	0	0	0
Threshers	0	1	0	0	0	0	2	1	0
Manual fodder cutter	0	3	11	2	2	2	5	6	1
Electrical fodder cutter	0	7	4	3	6	0	4	0	8
Diesel fodder cutter	0	1	1	7	0	0	2	1	0
Tractor+implements	0	0	1	2	2	0	1	0	2

Note: Village groups: 1=Godam, 2=Mandpa, 3=Sukhomajri, 4=Dhamala, 5=Main Nada, 9=Harijan Nada, 6=Masoompur, 7=Thathar, 8=Salehpur.

Table B.12. Irrigation facilities and village level indicators in Haryana.

Village group	1	2	3	4	5	9	6	7	8
6d. Irrigation facilities:									
Household getting water	0	10	17	9	9	4	10	1	6
From: dam	0	10	14	6	9	1	0	0	6
private tubewells	0	0	11	2	1	3	2	1	3
channel system	0	3	0	2	4	0	10	0	0
Number of dams	0	1	4	2	6	2	2	0	2
Number of operational dams	0	1	1	2	4	0	0	0	1
Catchment area (ha)	0	2	5	2	10	0	0	0	2
Command area (ha)	0	40	20	30	120	0	0	0	40
Per capita capital (Rs 000)	4	265	170	197	431	146	22	189	154
Household capital (Rs 000)	23	1446	1114	1470	3606	962	145	1498	1053
7. Village level indicators:									
Level of participation, based on:									
total sum	0.841	0.604	0.707	0.633	0.528	0.561	0.464	0.664	0.789
first principal component	0.698	0.621	0.542	0.501	0.610	0.716	0.546	0.780	0.729
second principal component	0.742	0.495	0.755	0.718	0.382	0.326	0.348	0.356	0.646
Years of protection	5.8	7.2	13.5	13.5	10.8	11.1	9.3	4.2	6.3
Forest dependence	0.882	0.609	0.649	0.582	0.762	0.505	0.649	0.786	0.658
Environmental interest	0.858	0.683	0.935	0.845	0.863	0.750	0.823	0.750	0.813
Level of resources (*R*)	0.603	0.385	0.668	0.649	0.622	0.481	0.647	0.484	0.815

Note: Village groups: 1=Godam, 2=Mandpa, 3=Sukhomajri, 4=Dhamala, 5=Main Nada, 9=Harijan Nada, 6=Masoompur, 7=Thathar, 8=Salehpur.

Table B.13. Demographics and inequalities in Jharkhand.

Village group	1	2	3	4	5	6	7	8	9	10	11	12
1. Demographics:												
Number of households	64	128	69	24	40	43	45	50	151	50	120	60
Averages in the field survey:												
Number of households	7	16	16	9	6	9	9	7	16	10	9	9
Number of family members	58	129	129	57	53	57	76	68	162	60	64	64
Number of women	29	71	68	31	25	23	38	40	79	34	30	29
Number of adults	30	61	66	32	33	31	43	33	90	36	36	40
Number of children	28	68	63	25	20	26	33	35	72	24	28	24
Female-male ratio	1000	1101	1054	1088	943	807	1000	1176	975	1133	938	906
Number of people employed	32	73	64	35	32	34	48	38	89	40	38	42
employed outside village	1	9	7	3	3	12	9	3	13	4	4	8
Years of schooling (age above 15)	3.56	3.53	4.39	1.90	5.86	1.85	3.58	3.14	4.91	3.42	3.57	1.47
2. Inequalities:												
2a. Caste:												
Brahmin	0	1	0	0	0	0	0	0	14	1	3	0
Rajput	1	2	0	0	0	0	0	0	0	0	0	0
Backward	2	1	0	0	1	1	3	3	1	1	2	3
Scheduled castes	0	0	0	0	0	0	1	1	0	1	3	1
Scheduled tribes	4	12	16	9	5	8	5	3	1	7	1	5
Equality of caste	0.429	0.602	1.000	1.000	0.715	0.811	0.505	0.388	0.790	0.530	0.316	0.505
2b. Religion:												
Hindu	7	13	3	6	3	9	8	6	16	10	9	9
Muslim	0	0	0	0	0	0	1	1	0	0	0	0
Christian	0	3	13	3	3	0	0	0	0	0	0	0
Equality of religion	1.000	0.695	0.695	0.556	0.500	1.000	0.802	0.755	1.000	1.000	1.000	1.000
Equality index	0.714	0.649	0.848	0.778	0.607	0.905	0.654	0.571	0.895	0.765	0.658	0.752
Gini-coefficient	0.071	0.207	0.054	0.046	0.119	0.087	0.014	0.047	0.200	0.207	0.088	0.080

Note: Village groups: 1=Hendehas, 2=Chapri, 3= Barhania, 4=Sindhorwa, 5=Muru, 6=Bhusaria, 7=Bakhari, 8=Kumbhawa, 9=Sakanpirhi and Tandwa, 10=Mundaria, 11= Kashia, 12=Khamdih; CVP = *chakriya vikas pranali*.

Table B.14. Classification of poolers and various types of income in Jharkhand.

Village group	1	2	3	4	5	6	7	8	9	10	11	12
3. Classification of poolers:												
No Pooler	0	4	4	8	0	3	6	0	2	5	3	1
Pooler	7	8	12	1	4	2	3	7	13	5	6	8
Withdrawer	0	4	0	0	2	4	0	0	1	0	0	0
4. Income: (per year)												
4a. Average household level income: (Rs)												
From land	3359	2177	-301	-90	3247	1853	-1228	3692	3993	-734	1215	560
Land income per ha	1580	909	-98	-137	1025	1573	-1328	2243	1732	-448	1424	828
From cattle	-218	-342	372	1008	457	56	933	175	353	-1939	-27	-229
Cattle income per cattle	-92	-107	71	231	115	50	294	30	88	-725	-16	-95
Outside village income	1800	10800	4088	7467	4200	6933	6953	2571	7600	7700	1737	4111
Within village without CVP	471	1525	71	3067	2000	1133	5240	1907	2550	3019	1478	400
Total household income	11664	17353	11260	12363	15379	9975	13775	12395	14496	12304	8603	10482
Per capita income	1357	2090	1462	1876	1945	1689	1668	1343	1362	1745	1199	1660
4b. Average output: (in quintals per year)												
Agriculture	48	234	163	46	180	67	108	156	707	87	141	61
Yield rate per acre	3.19	6.11	3.31	7.82	9.46	6.36	12.96	13.56	19.16	5.30	18.33	10.08
Animal husbandry	6	30	2	0	3	2	21	18	61	22	3	8
Yield rate per cattle	1.00	1.23	0.06	0.00	0.34	0.50	1.95	0.74	1.73	1.76	1.12	1.04
4c. Average benefits from participation: (Rs)												
Present value of the expected share from the plantation												
Student share	4085	334	851	60907	2567	340	165	276	4356	820	45	203
Land-pooler's share	4085	446	0	1238	2567	528	495	322	5361	2733	408	304
Income through CVP	6252	3192	7029	913	5475	0	1877	4050	0	4258	4200	5640
Total payoff	14422	3861	8731	61819	10610	679	2207	4602	8711	5898	4291	6046
4d. Benefits at the village level: (Rs)												
Village fund	9554	195709	12732	19400	22839	76180	10208	11441	54221	10626	10829	9392

Note: Village groups: 1=Hendehas, 2=Chapri, 3= Barhania, 4=Sindhorwa, 5=Muru, 6=Bhusaria, 7=Bakhari, 8=Kumbhawa, 9=Sakanpirhi and Tandwa, 10=Mundaria, 11= Kashia, 12=Khamdih; CVP = *chakriya vikas pranali*.

Table B.15. Consumption, land- and cattle-ownership in Jharkhand.

Village group	1	2	3	4	5	6	7	8	9	10	11	12
5. Consumption (per year)												
Number of families who let their cattle graze freely	1	11	16	7	4	4	6	6	5	8	6	7
Fodder per cattle (qu)	3.35	3.69	1.42	0.82	3.92	5.06	2.29	2.20	11.50	4.83	5.40	2.07
Fuelwood per househ (qu)	20.0	30.1	59.4	43.3	33.2	35.6	37.6	33.6	27.1	15.2	28.3	20.7
Fuelwood per capita (qu)	2.41	3.73	7.36	6.84	3.75	5.61	4.45	3.46	2.68	2.53	3.98	2.91
Food per household (Rs)	10719	13152	10148	12219	15743	892	13732	11774	15714	2657	10998	12172
Food per capita (Rs)	1408	1680	1290	1870	1988	1783	1666	1253	1467	1935	1499	1243
6. Capital:												
6a. Land-ownership												
Size of the pool (acres)	48	23	88	95	2	102	41	40	105	41	45	41
Accessible forest (acres)	0	0	3770	0	0	150	0	0	0	0	0	
Land owned (acres)	50	116	182	15	76	43	24	37	115	53	29	21
Land cultivated (acres)	37	96	123	15	48	27	21	29	92	41	19	15
Land irrigated (acres)	14	6	32	4	16	1	8	7	44	2	13	4
Number of landowners	7	16	16	6	6	5	9	7	16	10	8	9
6b. Cattle-ownership: (number)												
Buffaloes	0	3	0	0	0	0	4	1	16	5	0	1
Cows	5	13	19	18	6	2	3	18	16	5	2	6
Goats	2	31	65	29	11	8	11	22	0	7	2	2
Draught animals	11	27	48	14	15	6	18	16	29	14	13	14
Number of cattle owners	4	9	13	6	3	4	7	5	13	8	2	2
Number of milch animals	1.38	2.71	2.71	4.21	2.92	1.00	1.51	4.94	2.71	1.59	1.25	3.85

Note: Village groups: 1=Hendehas, 2=Chapri, 3= Barhania, 4=Sindhorwa, 5=Muru, 6=Bhusaria, 7=Bakhari, 8=Kumbhawa, 9=Sakanpirhi and Tandwa, 10=Mundaria, 11= Kashia, 12=Khamdih.

Table B.16. Private assets, irrigation facilities and village level indicators in Jharkhand.

Village group	1	2	3	4	5	6	7	8	9	10	11	12
6c. Private-assets:												
Wells	0	2	21	1	9	1	8	7	22	4	3	1
Pumps	0	0	1	0	3	1	1	1	10	2	0	0
Persian wheel	0	0	0	0	0	0	1	2	4	1	0	0
Ponds	0	0	2	0	2	0	0	0	0	0	0	0
Ox-cart	0	0	0	0	0	0	0	0	1	0	0	0
Cane crushers	0	0	0	0	0	0	0	0	1	0	0	0
Manual fodder cutter	0	0	0	0	0	0	0	0	2	1	0	0
6d. Irrigation facilities:												
Household getting water	7	4	13	4	5	1	8	7	16	2	7	3
From: channel system	1	0	0	0	0	0	0	0	0	0	7	2
Per capita capital (Rs 000)	12	30	47	13	36	0	11	25	34	3	19	12
Household capital (Rs 000)	95	215	376	77	327	0	88	211	321	7	142	74
7. Village level indicators												
Level of participation, based on:												
total sum	0.849	0.491	0.846	0.624	0.534	0.546	0.712	0.719	0.763	0.659	0.703	0.654
first principal component	0.831	0.727	0.807	0.825	0.761	0.875	0.857	0.800	0.860	0.781	0.837	0.822
second principal component	0.801	0.381	0.772	0.528	0.562	0.595	0.485	0.578	0.708	0.602	0.568	0.519
third principal component	0.602	0.632	0.692	0.639	0.419	0.352	0.730	0.711	0.578	0.597	0.636	0.607
pooling intensity	0.388	0.326	0.374	0.205	0.264	0.295	0.129	0.240	0.195	0.256	0.249	0.332
Years of protection	2	9	2	7	8	9	2	2	7	2	2	2
Environmental interest	0.965	0.845	0.970	0.973	0.708	0.695	0.945	0.750	0.938	0.925	0.890	0.890
Level of resources (*R*)	0.849	0.723	0.781	0.943	0.283	0.709	0.744	0.608	0.685	0.546	0.651	0.446

Note: Village groups: 1=Hendehas, 2=Chapri, 3= Barhania, 4=Sindhorwa, 5=Muru, 6=Bhusaria, 7=Bakhari, 8=Kumbhawa, 9=Sakanpirhi and Tandwa, 10=Mundaria, 11= Kashia, 12=Khamdih.

Table B.17. Demographics, inequalities, forest-ownership and average Output in Uttaranchal.

Village group	1	2	3	4	5	6	7	8	9	10	11
1. Demographics:											
Number of households	91	38	70	500	200	250	70	42	170	122	75
Averages in the field survey:											
Number of households	14	8	5	23	13	15	6	5	22	8	16
Number of family members	72	48	24	134	73	103	56	25	157	65	96
Number of women	33	23	13	68	37	50	27	12	76	34	50
Number of adults	41	25	15	82	44	52	26	14	91	32	58
Number of children	31	23	9	52	29	51	30	11	66	33	38
Female-male ratio	917	958	1083	1015	1014	971	964	960	968	1046	1042
Number of people employed	57	37	21	94	51	76	39	18	106	44	55
employed outside the village	10	4	6	18	9	9	5	3	20	6	7
Years of schooling (age above 15)	5.57	7.00	9.60	6.04	7.38	6.60	3.50	6.20	6.00	3.88	5.38
2. Inequalities:											
2a. Caste:											
Brahmin	0	1	0	0	4	4	0	0	1	0	0
Ksatriya: Rajput	11	6	5	22	9	7	6	5	20	8	15
Other (low caste)	3	1	0	1	0	4	0	0	1	0	1
Equality of caste	0.663	0.594	1.000	0.917	0.574	0.360	1.000	1.000	0.831	1.000	0.883
Gini-coefficient	0.255	0.133	0.212	0.206	0.153	0.1076	0.039	0.146	0.2185	0.168	0.266
3. Owner of the forest:											
Sarpanch	5	1	1	10	8	7	3	1	6	4	7
Villagers	6	4	4	10	5	7	2	4	15	4	6
Government	2	1	0	3	0	1	1	0	0	0	1
Other	1	2	0	0	0	0	0	0	1	0	2
4. Income: (per year)											
4b. Average output: (in quintals per year)											
Agriculture	91	53	39	106	213	174	63	55	826	345	229
Yield rate per ha	20.3	11.8	9.3	16.3	38.2	30.1	20.1	22.0	73.8	51.6	28.9
Animal husbandry	51	32	22	78	48	86	45	69	206	65	86
Yield rate per cattle	2.55	1.52	1.83	1.70	2.09	2.15	1.80	4.06	2.12	2.32	2.69

Note: Village groups: 1=Ulli, 2=Than, 3= Thamana, 4=Nishni, 5=Bachher, 6=Makkumath, 7=Ushara, 8=Sagar, 9=Shama, 10=Panyali, 11= Reema.

Table B.18. Income and benefits from participation in Uttaranchal.

Village group	1	2	3	4	5	6	7	8	9	10	11
4. Income: (per year)											
4a. Average household level income: (Rs)											
From land	-308	-378	1040	-713	-161	-69	686	350	6941	4353	33
Land income per acre	-963	-671	1238	-2515	-374	-179	1311	700	13635	5213	66
From cattle	546	2156	0	-151	638	380	1917	9620	1110	-169	-685
Cattle income per cattle	212	595	0	-48	193	89	329	1781	186	-31	-215
Outside village income	15379	16225	61240	18871	28400	8608	11100	15200	27200	12755	16338
Within village income	5531	4258	5840	3723	2909	7084	7603	15970	12870	10011	8180
Total household income	20910	20483	67080	22594	31309	15692	18703	31170	40070	22766	24517
Per capita income	6136	3865	15007	4398	5535	2254	2061	6361	7135	2870	4947
4c. Average benefits from participation: (in quintals)											
Fodder from:											
Reserved forest	15.5	1.9	18.0	3.2	16.7	0.0	14.2	0.0	0.9	0.0	7.4
communal forest	6.9	2.1	0.0	28.5	5.6	6.4	17.5	18.0	17.8	23.1	10.4
civil/soyam forest	0.0	22.5	6.0	0.3	5.0	34.5	30.0	55.0	0.0	10.0	0.0
Fuelwood from:											
Reserved forest	11.3	0.6	0.0	1.7	5.8	1.3	9.2	0.0	9.4	3.1	10.3
communal forest	3.1	1.9	0.0	9.7	5.2	8.4	19.2	12.0	27.7	33.8	8.4
civil/soyam forest	0.5	16.0	0.0	0.0	2.8	13.1	16.7	13.8	0.0	5.6	0.0
Timber from:											
Reserved forest	0.1	0.0	0.0	0.0	0.0	0.0	0.0	0.0	0.2	0.0	8.6
communal forest	0.1	0.0	0.0	0.0	0.6	0.3	1.0	4.8	5.3	4.3	0.0
civil/soyam forest	0.0	0.0	0.0	0.0	0.0	0.7	0.0	0.0	0.0	0.0	0.0
Payoff from: (Rs)											
Reserved forest	2339	245	2037	454	2187	69	2077	0	592	162	1479
communal forest	936	337	0	3724	910	1162	2984	2718	3517	4416	1618
civil/soyam forest	26	3374	679	34	709	4596	4257	6938	0	1423	0
Total payoff	3300	3956	2716	4213	3806	5827	9319	9656	4110	6000	3097
Village fund	0	0	0	0	12000	120000	0	0	15000	0	0

Note: Village groups: 1=Ulli, 2=Than, 3= Thamana, 4=Nishni, 5=Bachher, 6=Makkumath, 7=Ushara, 8=Sagar, 9=Shama, 10=Panyali, 11= Reema.

Table B.19. Consumption and capital in Uttaranchal.

Village group	1	2	3	4	5	6	7	8	9	10	11
5. Consumption (per year)											
Fodder per cattle (qu)	14.8	15.4	11.9	15.6	13.9	14.9	13.9	16.3	11.7	16.2	17.0
Fuelwood per household (qu)	14.9	18.5	9.0	11.7	14.0	23.7	45.0	25.8	41.6	45.0	21.2
Fuelwood per capita (qu)	2.90	3.08	1.88	2.01	2.49	3.45	4.82	5.16	5.83	5.54	3.53
Food per household (Rs)	16722	18478	30723	19767	24603	14608	19362	23565	27478	23356	14633
Food per capita (Rs)	4108	3351	6952	3694	4799	2168	2157	4783	4807	2980	2662
6. Capital:											
Accessible common forests: (in ha)											
Communal forest	60	6	10	1000	182	2200	1016		680	50	28
Civil/soyam forest	0	50	0	0	182	200	400		0	40	0
Reserved forest	1500	0	100	0	0	0	208		0	60	60
6a. Land-ownership:											
Land owned (nalis)	309	225	450	359	329	312	157	125	699	334	450
Land cultivated (nalis)	224	225	210	326	279	289	157	125	560	334	396
Land irrigated (nalis)	15	8	25	5	49	20	0	13	161	2	175
Number of landowners	14	8	5	23	13	15	6	5	21	8	16
6b. Cattle-ownership: (number)											
Buffaloes	8	6	0	18	7	9	7	7	15	9	15
Cows	9	11	12	24	15	30	12	9	57	18	12
Goats	9	13	0	5	0	0	20	0	88	0	8
Draught animals	16	8	4	26	20	24	10	10	34	16	19
Number of cattle owners	13	8	4	22	12	15	6	5	21	8	15
Average number of milch animals per household	1.43	2.63	3.00	2.09	1.92	2.67	4.17	3.40	4.62	3.50	2.13

Note: Village groups: 1=Ulli, 2=Than, 3= Thamana, 4=Nishni, 5=Bachher, 6=Makkumath, 7=Ushara, 8=Sagar, 9=Shama, 10=Panyali, 11= Reema.

Table B.20. Irrigation facilities and village level indicators in Uttaranchal.

Village group	1	2	3	4	5	6	7	8	9	10	11
6d. Irrigation facilities:											
Households getting water	8	3	3	1	8	2	0	4	14	1	15
From: channel	2	1	1	0	3	0	0	4	9	1	10
river	6	2	0	0	3	2	0	0	0	0	4
tap	1	0	2	1	2	0	0	0	5	0	1
Per capita capital (Rs 000)	54	51	123	30	59	36	32	72	52	26	38
Household capital (Rs 000)	176	273	435	166	211	244	217	348	311	221	213
7. Village level indicators:											
Level of participation, based on:											
total sum	0.635	0.646	0.240	0.590	0.635	0.621	0.738	0.583	0.595	0.632	0.523
first principal component	0.728	0.735	0.316	0.702	0.670	0.724	0.781	0.654	0.665	0.695	0.619
second principal component	0.464	0.475	0.457	0.421	0.493	0.454	0.603	0.502	0.479	0.513	0.430
third principal component	0.551	0.568	0.609	0.504	0.432	0.400	0.241	0.489	0.472	0.354	0.403
fourth principal component	0.566	0.489	0.309	0.490	0.657	0.615	0.590	0.516	0.555	0.603	0.607
Registration date of *van panchayat*	1948		1936	1943	1980	1956	1965	1965			
Forest dependence	0.834	0.741	0.231	0.819	0.763	0.801	0.884	0.930	0.658	0.790	0.638
Environmental interest	0.928	0.970	1.000	0.925	1.000	0.850	0.958	0.900	0.840	0.875	0.860
Level of Resources (*R*)	0.434	0.612	0.174	0.417	0.890	0.482	0.604	0.423	0.422	0.630	0.677
Type of guard(s)	ft	rs	no	ft/rs	ft	ft/ft	ft/rs	no	ft/no	ft	ft

ft = full time; rs = rotational system; no = not anymore.

Note: Village groups: 1=Ulli, 2=Than, 3= Thamana, 4=Nishni, 5=Bachher, 6=Makkumath, 7=Ushara, 8=Sagar, 9=Shama, 10=Panyali, 11= Reema.

References

[1] D. Abreu, D. Pearce, and E. Stacchetti. Toward a theory of discounted repeated games with imperfect monitoring. *Econometrica*, **58**:1041–1064, 1990.

[2] C. Agarwal. Boundary and property rights in Uttarakhand forests. *Wasteland News*, **11**(1):4–6, 1996.

[3] A. A. Alchian. Uncertainty, evolution and economic theory. *Journal of Political Economy*, **58**(3):211–221, 1950.

[4] A. Ali, N. Bajpai, M. Halan, B. Joshi, and U. Narain. Project to study the interrelationship between forestry and agriculture and the institution of van panchayats in Almora district of the U.P.hills. Delhi School of Economics and Centre for Science and Education, 1990.

[5] E. Amundsen and L. Bergman. Will cross-ownership re-establish market power in the nordic power market? *The Energy Journal*, **23**(2):73–95, 2002.

[6] T. C. A. Anant. Property rights and natural resources. *Environmental Economics*, pages 36–50, 1992.

[7] H. Arora and A. Khare. Experience with the recent Joint Forest Management approach. Society for Promotion of Wastelands Development, New Delhi, Feb. 1994. Paper prepared for the international workshop on India's forest management and ecological revival.

[8] M. Ashish, S. L. Shah, and M. G. Jackson. Forest panchayats. Resource management by the people, for the people. Personal communication, 1990.

[9] R. Axelrod. *The Evolution of Co-operation*. Penguin books, London, 1984.

[10] P. Bardhan. Analytics of the institutions of informal cooperation in rural development. *World Development*, **21**(4):633–639, 1993.

[11] P. Bardhan. Rational fools and cooperation in a poor hydraulic economy. In K. Basu, P. Pattanaik, and K. Suzumura, editors, *Choice, Welfare, and Development, A Festschrift in Honour of Amartya K. Sen*. Oxford University press, Oxford, 1995.

[12] K. Basu. *The Less Developed Economy, a Critique of Contemporary Theory*. Oxford University press, Oxford, 1984.

[13] K. Basu. Modelling finitely-repeated games with uncertain termination. *Economic Letters*, **23**:147–151, 1987.

[14] K. Basu and A. Mishra. Sustainable development and the commons problem, a simple approach. In P. Bardhan, M. D. Chaudhuri, and T. N. Krishnan, editors, *Development and Change: Essays in Honour of K. N. Raj*, chapter 6. Oxford University press, Bombay, 1993.

[15] R. H. Bates. Contra contractarianism: some reflections on the new institutionalism. *Politics and Society*, **16**:387–401, 1988.

[16] B. Bayramoğlu. *Climate Change and International Institutions: Agents of Global Environmental Cooperation*. PhD thesis, Department of International Relations, Bilkent University, Ankara, Turkey, December 1997.

[17] B. Bayramoğlu-Lise and W. Lise. Climate change, environmental NGOs and public awareness in the Netherlands: Perceptions and reality. *The ICFAI Journal of Environmental Law*, **III**(1):55–67, 2004.

[18] H. Behrendt, A. Gilbert, J. Hofmann, R. Janssen, A. Kannen, H. Lenhart, W. Lise, C. Nunneri, and W. Windhorst. Catchment-coastal zone interaction based upon scenario and model analysis: case study Elbe and the German Bight. *Regional Environmental Change*, **5**:54–81, 2005.

[19] J. Bendor and D. Mookherjee. Institutional structure and the logic of ongoing collective action. *American Political Science Review*, **81**(1):129–154, 1987.

[20] A. S. Bist. *'Van jage vanvasi jage' (Forest people go where the forests go)*. Uttarakhand seva nidhi, Almora, 1993.

[21] A. S. Bist and H. Gairola, editors. *'Van prabandhan evan van panchayat sammelan, Dogri kandai (janpad Chamoli), paricharcha prarup' (Conference on forest management and van panchayats, Dogri section (Chamoli district))*, Gopeshwar, Mar. 1994. Jakeshwar shikshan sansthan.

[22] T. F. Bresnahan. Empirical studies of industries with market power. In R. Schmalensee and R. D. Willig, editors, *Handbook of industrial Organization, Volume II*. Elsevier Science Publishers, Amsterdam, 1989.

[23] C. Britt. Community, committees and common forests: a history of local institutions and forest management in two central Himalayan villages. Discussion paper, International Centre for Integrated Mountain Development, 1993.

[24] C. Britt. Nature of women's participation in local institutions and common forest management. *Wasteland News*, **XI**(1):59–64, 1995.

[25] D. W. Bromley. *Environment and Economy: Property Rights and Public Policy*. Basil Blackwell, Oxford, 1991.

[26] D. W. Bromley. Testing for common versus private property: comment. *Journal of Environmental Economics and Management*, 21:92–96, 1991.

[27] D. W. Bromley and M. M. Cernea. *The Management of Common Property Natural Resources: Some Conceptual and Operational Fallacities*. The World Bank, Washington, 1989.

[28] S. Burkey. *People First, a Guide to Self-reliant, Participatory Rural Development*. Zed-books, London, 1993.

[29] E. N. Castle. A conceptual framework for the study of rural places. *American Journal of Agricultural Economics*, **80**:621–631, 1998.

[30] M. Chakraborty. An analysis of the causes of deforestation in India. In K. Brown and D. W. Pearce, editors, *The Causes of Tropical Deforestation, The economic and statistical analysis of factors giving rise to the loss of the tropical forests*, chapter 16. UCL press, London, 1995.

[31] K. Chopra, G. K. Kadekodi, and M. N. Murty. *Participatory Development, People and Common Property Resources*. Sage publications, New Delhi, 1990.

[32] J. Choubey and R. M. Mehta. Barhania story, land financed school through Chakriya Vikas Pranali. Society of Hill Resource Management School, Daltenganj, 1995.

[33] S. V. Ciriacy-Wantrup and R. C. Bishop. Common property as a concept in natural resource management. *Natural Resources Journal*, **15**:713–727, 1975.

[34] R. H. Coase. The problem of social cost. *Journal of Law and Economics*, **3**:1–44, 1960.

[35] R. H. Coase. The new institutional economics. *Journal of Institutional and Theoretical Economics*, **140**(1):229–231, 1984.

[36] R. H. Coase. The institutional structure of production. *American Economic Review*, **82**(4):713–719, 1992.

[37] J. S. Coleman. *Foundations of Social Theory*. Harvard University Press, Harvard, 1990.

[38] J. R. Commons. *The Economics of Collective Action*. Macmillan, New York, 1950.

[39] J. R. Commons. *Institutional Economics*. University of Wisconsin press, Madison, WI, 1961.

[40] B. R. Copeland and M. S. Taylor. Trade, tragedy, and the commons. NBER Working Paper W10836, National Bureau of Economic Research, Cambridge, USA, 2004.

[41] L. de Alessi. The economics of property rights: a review of the evidence. *Research in Law and Economics*, **2**:1–47, 1980.

[42] P. M. DeMarzo. Coalition, leadership and social norms: the power of suggestion in games. *Games and Economic Behaviour*, **4**(1):72–100, 1992.

[43] S. K. Dhar. Rehabilitation of degraded tropical forest watershed with people's participation. Tata Energy Research Institute, 1994. *Joint Forest Management Series* **16**.

[44] S. K. Dhar, K. Shah, and M. Sarin. *Participatory forest management in the Shivalik hills: experiences of the Haryana forest department*. Ford Foundation, New Delhi, 1990. Working paper No. 5.

[45] T. Dietz, E. Ostrom, and P. C. Stern. The struggle to govern the commons: tragedy of the commons? *Science*, **302**:1907–1912, 2003.

[46] W. Dugger. The new institutionalism: new but not institutionalist. *Journal of Economic Issues*, **24**(2):423–431, 1990.

[47] A. Duraiappah. Poverty and environmental degradation: a review and analysis of the nexus. *World Development*, **26**(12):2169–2179, 1998.

[48] T. Eggertsson. *Economic Behaviour and Institutions*. Cambridge University press, New York, 1990.

[49] T. Eggertsson. The economics of institutions, avoiding the open-field syndrome and the perils of path dependence. *Acta Sociologica*, **36**(3):227–237, 1993.

[50] J. Ensminger and A. Rutten. The political economy of changing property rights, dismantling a pastoral commons. *American Ethnologist*, **18**(4):683–699, 1991.

[51] A. J. Field. Microeconomics, norms and rationality. *Economic Development and Cultural Change*, **32**:683–711, 1984.

[52] B. C. Field. The evolution of property rights. *Kyklos*, **42**(3):319–345, 1989.

[53] J. W. Friedman. Cooperative equilibria in finite horizon noncooperative supergames. *Journal of Economic Theory*, **35**:390–398, 1985.

[54] N. Frohlich. An impartial reasoning solution to the prisoner's dilemma. *Public Choice*, **74**:447–460, 1992.

[55] D. Fudenberg and E. Maskin. The Folk theorem in repeated games with discounting or with incomplete information. *Econometrica*, **54**:533–554, 1986.

[56] D. Fudenberg and E. Maskin. Evolution and cooperation in noisy repeated games. *American Economic Review*, **80**(2):274–279, May 1990.

[57] F. Fukuyama. *Trust: Social Virtues and the Creation Of*. Free Press, New York, 1995.

[58] E. G. Furubotn and S. Pejovich. Property rights and economic theory: a survey of recent literature. *Journal of Economic Literature*, **10**:1137–1162, 1972.

[59] E. G. Furubotn and R. Richter. The new institutional economics. The new institutional approach to economic history. *Journal of Institutional and Theoretical Economics*, **145**:1–245, 1989.

[60] E. G. Furubotn and R. Richter. *The New Institutional Economics*. J. C. Mohr, Tubingen, 1991.

[61] M. Gadgil and R. Guha. Ecological conflicts and the environmental movement in India. *Development and Change*, **25**(1):101–136, 1994.

[62] R. Gardner, E. Ostrom, and J. M. Walker. The nature of the common-pool resource problem. *Rationality and Society*, **2**:335–358, 1990.

[63] D. Genesove. Search at wholesale auto auctions. *The Quarterly Journal of Economics*, **109**:23–49, 1995.

[64] R. Gerlagh and W. Lise. Carbon taxes: a drop in the ocean, or a drop that erodes the stone? the effect of carbon taxes on technological change. *Ecological Economics*, **54**(2–3):241–260, 2005.

[65] R. Gibbons. An introduction to applicable game theory. *Journal of Economic Perspectives*, **11**(1):127–149, 1997.

[66] C. Ginasante, M. Aguilar, L. Babiano, A. Garrido, A. Gmez, E. Iglesias, W. Lise, L. Moral, and B. Pedregal. Institutional adaptation to changing risk of water scarcity in the Lower Guadalquivir basin. *Natural Resources Journal*, **42**(3):521–563, 2002.

[67] E. J. Green and R. H. Porter. Noncooperative collusion under imperfect price information. *Econometrica*, **52**(1):87–100, 1984.

[68] J. Greenberg. *The Theory of Social Situations, an Alternative Game-theoretic Approach*. University press, Cambridge, 1st edition, 1990.

[69] A. Greif, P. Milgrom, and B. R. Weingast. Coordination, commitment, and enforcement: The case of the merchant guild. *Journal of Political Economy*, **102**(4):745–776, 1994.

[70] R. Guha. Forestry in British and post-British India, a historical analysis. *Economic and Political Weekly*, **18**:1882–1896; 1940–1947, 1983.

[71] R. Guha. Scientific forestry and social change in Uttarakhand. *Economic and Political Weekly*, **20**(45, 46 and 47):1939–1952, 1985.

[72] R. Guha. Dietrich Brandis and Indian forestry: A vision revisited and reaffirmed. In M. Poffenberger and B. McGean, editors, *Village Voices, Forest Choices, Joint Forest Management in India*, chapter 3. Oxford University press, Delhi, 1996.

[73] J. R. Gupta. Evolution of a legal agreement for joint forest management in Haryana. *Wasteland News*, **10**(3):38–48, 1995.

[74] G. Hardin. The tragedy of the commons. *Science*, **162**:1243–1248, 1968.

[75] G. Hardin and J. Baden. *Managing the Commons*. Freeman publishing Co., San Francisco, 1977.

[76] R. Hardin. Collective action as an agreeable n-person prisoner's dilemma. *Behavioural Science*, **16**:472–481, 1971.

[77] R. Hardin. *Collective Action*. John Hopkins University press, Resources for the future, Inc., Baltimore and London, 1982.

[78] H. H. Harman. *Modern Factor Analysis*. The university of Chicago press, Chicago, second revised edition, 1967.

[79] Haryana Forest Department and Tata Energy Research Institute. *Training workshop on joint/participatory forest management for block officers/deputy rangers of Haryana forest department*, Forest complex, Pinjore, 1993.

[80] Haryana Forest Department and Tata Energy Research Institute. *Training workshop on joint participatory forest management for members of management committees of Hill Resource Management Societies*, Forest complex, Pinjore, 1994.

[81] Y. Hayami and M. Kikuchi. *Asian Village Economy at the Crossroads: an Economic Approach to Institutional Change*. University of Tokyo press, Tokyo, 1981.

[82] HFD. Joint Forest Management policy and rules (in hindi). Haryana Forest Department, Deputy Conservator of Forests, Research and Training, Forest Division, Pinjore, 1993.

[83] D. Hirschleifer and E. Rasmusen. Cooperation in a repeated prisoner's dilemma with ostracism. *Journal of Economic Behaviour and Organisation*, **12**:87–106, 1989.

[84] L. Hotte, N. V. Long, and H. Tian. International trade with endogenous enforcement of property rights. *Journal of Development Economics*, **62**:25–54, 2000.

[85] M. C. W. Janssen, J. Gorter, and S. van de Meerendonk. Cooperation in a modified version of the finitely repeated prisoners' dilemma game. *Journal of Economic Behavior & Organization*, **32**(4):613–619, Apr. 1997.

[86] G. K. Kadekodi. Cycle of change. *Down to earth*, pages 27–32, August 31 1995.

[87] G. K. Kadekodi. Operationalising sustainable development, ecology-economy interactions at regional level, case of Palamau district. Technical report, Institute of Economic Growth and Institute for Environmental Studies, Delhi and Amsterdam, Mar. 1995.

[88] G. K. Kadekodi and K. Chopra. Cyclic system of development, a monograph. Institute of Economic Growth, Delhi, 1989.

[89] G. K. Kadekodi and K. Chopra. Cyclical reinvestment strategy: an alternative approach to rural development. *Social Action*, **40**(4):382–397, 1990.

[90] G. K. Kadekodi and W. Lise. Investment on ecology with people's participation. *Pacific and Asian Journal of Energy*, **4**(2):89–98, 1995.

[91] J. H. Kagel and A. E. Roth, editors. *The handbook of experimental economics*. Princeton University Press, Princeton, 1995.

[92] M. Kandori. Social norms and community enforcement. *Review of Economic Studies*, **59**(1):63–80, 1992.

[93] R. S. Kanetkar and V. Varalakshmi. Women in Godam - Haryana: a gender and caste based study on conservation of forest resources. Tata Energy Research Institute, 1994. *Joint Forest Management Series* **13**.

[94] O. N. Kaul, B. N. Dwivedi, and V. Shah. Ecological revival of Haryana Shivaliks through community participation. Tata Energy Research Institute, New Delhi, 1995. *Joint Forest Management Series* **17**.

[95] C. Kemfert, W. Lise, and R. S. J. Tol. Games of climate change with international trade. *Environmental and Resource Economics*, **28**(2):209–232, 2004.

[96] C. Keser and R. Gardner. Strategic behavior of experienced subjects in a common pool resource game. *International Journal of Game Theory*, **28**:241–252, 1999.

[97] L. R. Klein. Estimation of interdependent systems in macroeconometrics. *Econometrica*, **37**(2):171–192, 1969.

[98] T. Kloek and L. B. Mennes. Simultaneous equation estimation based on principal components of predetermined variables. *Econometrica*, **28**:45–61, 1960.

[99] T. C. Koopmans. *Three essays on the state of economic science*. McGraw-Hill, New York, 1957.

[100] D. M. Kreps, P. R. Milgrom, J. Roberts, and R. Wilson. Rational cooperation in the finitely repeated prisoner's dilemma. *Journal of Economic Theory*, **27**:245–252, 1982.

[101] D. M. Kreps and R. Wilson. Sequential equilibria. *Econometrica*, **50**:863–894, 1982.

[102] R. Kumar. Integrated land management - scarcity to surplus. Society for Promotion of Wastelands Development, New Delhi, 1993.

[103] C. Kunwar. *'Uttarakhand ki sablayen' (The Women of Uttarakhand)*. Himalayan Action Research Centre and Society for Participatory Research In Asia, Dehradun and Delhi, 1994.

[104] S. S. Kunwar. *Hugging the Himalayas, the Chipko Experience*. Dasholi Gram Swarajya Mandal, Gopeshwar, 1988.

[105] J. J. Laffont. Game theory and empirical economics: the case of auction data. *European Economic Review*, **41**:1–35, 1997.

[106] J. B. Lal. *India's Forests, Myth & Reality*. Natraj publishers, Dehra Dun, 1989.

[107] B. A. Larson and D. W. Bromley. Property rights, externalities and resource degradation, locating the tragedy. *Journal of Development Economics*, **33**:235–262, 1990.

[108] W. Lise. Socio-economic planning and project planning, India's experience of four decades. Planning Commission, New Delhi, practical training report prepared for Twente University, the Netherlands, Mar. 1991.

[109] W. Lise. A model of participatory development, with applications to rural India. Master's thesis, University of Twente, the Netherlands, 1992.

[110] W. Lise. Participatory development, a path to ecological preservation, an approach with case studies and modelling. *International Journal of Environment and Pollution*, 5(2/3):243–259, 1995.

[111] W. Lise. Preservation of the commons by pooling resources, modelled as a repeated game. Working paper No. 26, Centre for Development Economics, Delhi School of Economics, University Enclave, Delhi, Jan. 1995.

[112] W. Lise. *An Econometric and Game Theoretic Approach to Common Pool Resource Management: Case Studies in Rural India*. PhD thesis, Delhi School of Economics, University of Delhi, India, Aug. 1997.

[113] W. Lise. Linking indicators of participation in forest management and additional incomes from participating: An estimation procedure for non-cooperative games. Working paper No. E/184/97, Institute of Economic Growth, University Enclave, Delhi, Jan. 1997.

[114] W. Lise. A village betrayed. *Down to Earth*, **5**(21):38–39, March 31 1997.

[115] W. Lise. Factors influencing people's participation in forest management in India. *Ecological Economics*, **34**(3):379–392, 2000.

[116] W. Lise. Estimating a game theoretic model. *Computational Economics*, **18**(2):141–157, 2001.

[117] W. Lise. The CENER model: Cost-Effective Nutrient Emission Reduction (CENER) of the load to the North Sea from the Rhine and Elbe basin. Internally reviewed Report R-03/07, Institute for Environmental Studies, Free University, Amsterdam, 2003.

[118] W. Lise. The Dutch electricity market does liberalization bring security of supply? In A. G. Paşamehmetoğlu, A. Özgenoğlu, and A. Y. Yeşilay, editors, *Proceedings of the eight international symposium on Environmental Issues and Waste Management in Energy and Mineral Production*, SWEMP 2004, Kemer, Antalya, Turkey, 17-20 May 2004, pages 9–15. Atılım University, Ankara, Turkey, 2004.

[119] W. Lise. A game model of people's participation in forest management in northern India. *Environment and Development Economics*, **10**(2):217–240, 2005.

[120] W. Lise and K. Bakker. Economic regulation of the water supply industry in the UK: A game theoretic consideration of the implications for responding to drought risk. *International Journal of Water*, **3**(1):17–37, 2005.

[121] W. Lise, A. Garrido, and E. Iglesias. A game model of farmers' demand for irrigation water from reservoirs in southern Spain. *Risk Decision and Policy*, **6**(3):167–185, 2001.

[122] W. Lise, C. Kemfert, and R. S. J. Tol. The German electricity market - does liberalisation bring competition? *Nota di Lavoro* **3.03**, FEEM, Milan, Italy, 2003.

[123] W. Lise and V. Linderhof. Electricity market liberalisation in Europe - Who's got the power? Internally reviewed Report R-04/03, Institute for Environmental Studies, Free University, Amsterdam, 2004.

[124] W. Lise and R. S. J. Tol. Impact of climate on tourist demand. *Climatic Change*, **55**(4):429–449, 2002.

[125] W. Lise and R. S. J. Tol. Sensitivity of tourist destination choice to climate. In C. Giupponi and M. Shechter, editors, *Climate Change in the Mediterranean: Socio-economic Perspectives of Impacts, Vulnerability and Adaptation*, FEEM Series on Economics and the Environment, chapter 14. Edward Elgar Publishing Limited, Cheltenham, UK, 2003.

[126] W. Lise and R. S. J. Tol. Attainability of international environmental agreements as a social situation. *International Environmental Agreements: Politics, Law and Economics*, **4**(3):253–277, 2004.

[127] W. Lise and R. J. H. M. van der Veeren. Cost-effective nutrient emission reductions in the Rhine river basin. *Integrated Assessment: An International Journal*, **3**(4):321–342, 2002.

[128] D. Luce and H. Raiffa. *Games and Decisions, Introduction and critical survey*, chapter 5. Wiley, New York, 1957.

[129] M. Majumdar. Some approximation theorems on efficiency prices for infinite programs. *Journal of Economic Theory*, **2**:399–410, 1970.

[130] M. Majumdar. Decentralisation in infinite horizon economies: An introduction. *Journal of Economic Theory*, **45**:217–227, 1988.

[131] E. Malinvaud. Capital accumulation and efficient allocation of resources. *Econometrica*, **21**:233–268, 1953.

[132] H. M. Markowitz. *Portfolio Selection: Efficient Diversification of Investment*. Basil Blackwell, Cambridge, 2nd edition, 1991.

[133] H. Matsushima. Long-term partnership in a repeated prisoner's dilemma with random matching. *Economic Letters*, **34**:245–248, 1990.

[134] P. R. Milgrom, D. C. North, and B. R. Weingast. The role of institutions in the revival of trade: the law merchant, private judges and the champagne fairs. *Economics and Politics*, **2**(1):1–23, 1990.

[135] P. R. Milgrom and D. J. Roberts. Predation, reputation and the entry deterrence. *Journal of Economic Theory*, **27**:280–312, 1982.

[136] P. Mirowski. Is there a mathematical neo institutional economics? *Journal of Economic Issues*, **15**:593–613, 1981.

[137] P. R. Mishra. People's protection programme through *chakriya vikas pranali*, a case study. Ministry of Rural Development, Department of Wasteland Development, Government of India press, Minto Road, 1994.

[138] P. R. Mishra. 'Amar-sandesh, daula ka Sukhomajri' (Immortal tidings, the wealth of Sukhomajri). Society for Hill Resource Management School, Daltenganj, 1996, reprint.

[139] P. R. Mishra and M. Sarin. Social security through social fencing: Sukhomajri and Nada's road to self-sustaining development. Personal communication, Apr. 1987.

[140] P. R. Mishra and S. P. Sinha. Self reliant system for the masses through chakriya vikas pranali. Society for Hill Resource Management School, Daltenganj, 1994.

[141] M. Nishijima. Bargaining and social norms. Yokohama city University. Personal communication, 1991.

[142] D. C. North. The new institutional economics. *Journal of Institutional and Theoretical Economics*, **142**(1):230–237, 1986.

[143] D. C. North. Institutions, transaction cost and economic growth. *Economic Inquiry*, **25**:419–428, 1987.

[144] D. C. North. *Institutions, Institutional Change and Economic Performance*. Political Economy of Institutions and Decisions. Cambridge University press, Cambridge, 1990.

[145] D. C. North. "Institutions". *Journal of Economic Perspectives*, **5**(1):97–112, 1991.

[146] M. J. Norusis. *SPSS/PC+ Statistics 4.0 for the IBM PC/XT/AT and PS/2*. SPSS Inc., Chicago, 1990.

[147] A. Okada. The possibility of cooperation in an n-person prisoner's dilemma with institutional arrangements. *Public Choice*, **77**(3):629–656, 1993.

[148] M. Okuno-Fujiwara and A. Postlewaite. Social norms and random matching games. *Games and Economic Behavior*, **9**(1):79–109, 1995.

[149] E. Ostrom. An agenda for the study of institutions. *Public Choice*, **48**:3–25, 1986.

[150] E. Ostrom. *Governing the Commons, The Evolution of Institutions for Collective Action*. Political Economy of Institutions and Decisions. Cambridge University press, Cambridge, 1990.

[151] E. Ostrom and R. Gardner. Coping with asymmetries in the commons: self-governing irrigation systems can work. *Journal of Economic Perspectives*, **7**(4):93–112, 1993.

[152] E. Ostrom, R. Gardner, and J. Walker. *Rules, Games and Common-pool Resources*. Ann Arbor, The University of Michigan press, Michigan, 1994.

[153] E. Ostrom, J. Walker, and R. Gardner. Covenant with and without a sword, self-governance is possible. *American Political Science Review*, **86**(2):404–417, 1992.

[154] S. Palit. The future of Indian forest management: into the twenty-first century. Ford Foundation, New Delhi, Dec. 1993. Joint Forest Management working paper 15.

[155] A. Panda, M. Dabas, V. Varalakshmi, V. N. Shah, and J. R. Gupta. Grass yield under community participation in Haryana Shivaliks. Tata Energy Research Institute, 1992. *Joint Forest Management Series* **1**.

[156] A. Panda, V. Varalakshmi, J. R. Gupta, B. Singh, and M. Dabas. Impact of participatory forest management on the ecology of Shivalik hills in Haryana state. Tata Energy Research Institute, 1992. *Joint Forest Management Series* **2**.

[157] R. K. Pandey. *Van Panchayat*. Rajhans press, Tallital, Nainital, 1994.

[158] G. B. Pant. *Forest Problems in Kumaun. Forest Problems and National Uprisings in the Himalayan Region*. Gyanodaya Prakashan, Nainital, 1922. This was a reprint with a foreword of A. S. Rawat.

[159] P. O. Pineau and P. Murto. An oligopolistic investment model of the Finnish electricity market. *Annals of Operations Research*, **121**:123–148, 2003.

[160] M. Poffenberger and B. McGean, editors. *Village Voices, Forest Choices, Joint Forest Management in India*. Oxford University press, Delhi, 1996.

[161] M. Poffenberger and C. Singh. Emerging directions in Indian forest policy: legal framework for joint management. *Wasteland News*, Feb–Apr:4–11, 1992.

[162] S. Pongquan. *Participatory Development Activities at Local Level: Case Studies in Villages of Central Thailand: an Analysis of People's Participation in the Planned Development Process Using the Concept of the "Linking Loops"*. PhD thesis, Agricultural University Wageningen, the Netherlands, 1992.

[163] R. H. Porter. Optimal cartel trigger price strategies. *Journal of Economic Theory*, **29**:313–338, 1983.

[164] A. Poteete and E. Ostrom. An institutional approach to the study of forest resources. Technical report, International Forestry Resources and Institutions (IFRI) Research Program Indiana University, 2002. Available through http://www.cbnrm.net/pdf/poteete_a_001.pdf.

[165] B. P. Purohit. Forest panchayats in Uttarakhand and their administration in the context of the settlement of Chamoli. *Wasteland News*, **XII**(2):28–31, 1997.

[166] S. Qasim. Chipko's active torch-bearers. *Wasteland News*, **1**(3):9–10, 1985.

[167] A. Rapoport and M. Guyer. A taxonomy of 2x2 games. *General Systems Yearbook*, **11**:203–214, 1966.

[168] J. J. Rotemberg and G. Saloner. A supergame-theoretic model of price wars during booms. *American Economic Review*, **76**(3):390–407, 1986.

[169] C. F. Runge. Common property externalities: isolation, assurance and resource depletion in a traditional grazing context. *American Journal of Agricultural Economics*, **63**:595–606, 1981.

[170] C. F. Runge. Institutions and the free rider: the assurance problem in collective action. *The Journal of Politics*, **46**:154–181, 1984.

[171] C. F. Runge. Common property and collective action in economic development. *World Development*, **14**:623–635, 1986.

[172] M. Rutherford. *Institutions in Economics, the Old and the New Institutionalism*. Historical Perspectives on Modern Economics. Cambridge University press, Cambridge, 1994.

[173] V. W. Ruttan and Y. Hayami. Towards a theory of induced institutional innovation. *Journal of Development Studies*, **20**(4):203–223, 1984.

[174] L. Samuelson. A note on uncertainty and cooperation in a finitely repeated prisoner's dilemma. *International Journal of Game Theory*, **16**(3):187–195, 1987.

[175] M. Sarin. From conflict to collaboration: local institutions in Joint Forest Management. National Support Group for Joint Forest Management, Society for Promotion of Wastelands Development and the Ford Foundation, New Delhi, Nov. 1993. Joint Forest Management, *working paper* **14**.

[176] M. Sarin. JFM in Haryana. *Wasteland News*, **8**(2):3–4, 1993.

[177] M. Sarin. *Joint Forest Management The Haryana Experience*. Environment and Development Series. Centre for Environment Education, Ahmedabad, 1996.

[178] N. C. Saxena. *Towards Sustainable Forestry in the U. P. Hills*. Centre for Sustainable Development, Lal Bahadur Shastri National Academy of Administration, Mussoorie – 248179, 1996.

[179] E. Schlager and E. Ostrom. Property right regimes and natural resources: a conceptual analysis. *Land Economics*, **68**(3):249–262, 1992.

[180] A. Schotter. *The Economic Theory of Social Institutions*. Cambridge University press, Cambridge, 1981.

[181] P. Seabright. Managing local commons: theoretical issues in incentive design. *Journal of Economic Perspectives*, **7**(4):113–134, 1993.

[182] N. Sengupta. *Managing Common Property, Irrigation in India and the Philippines*. Sage, New Delhi, 1991.

[183] S. Sengupta. A few unresolved issues in Haryana. *Wasteland News*, **8**(2):40–45, 1993.

[184] R. Sethi and E. Somanathan. The evolution of social norms in common property use. *American Economic Review*, **86**(4):766–788, 1994.

[185] S. L. Shah. Functioning of van panchayats in 8 hill districts of Uttar Pradesh - an analysis of present malaise and lessons for the future in the context of the proposed van panchayat niyamawali 1989. Personal communication, 1990.

[186] SHRMS. Annual activities report of SHRMS 1989-90. Daltenganj, 1989.

[187] SHRMS. Annual activities report of SHRMS 1990-91. Daltenganj, 1990.

[188] SHRMS. *Annual Activities Report of SHRMS 1991*. Metro printing works, Kund Muhalla, Daltenganj, 1991.

[189] SHRMS. Annual activities report of SHRMS 1995-96. Daltenganj, 1996.

[190] P. Shyamsundar and R. A. Kramer. Tropical forest protection: An empirical analysis of the costs borne by local people. *Journal of Environmental Economics and Management*, **31**(2):129–144, 1996.

[191] B. Singh, S. S. Arora, and J. R. Gupta. Management of bamboo forest under joint participatory forest management in Haryana Shivaliks. Tata Energy Research Institute, 1993. *Joint Forest Management Series* **6**.

[192] K. Singh. *Managing Common Pool Resources, Principles and Case Studies*. Oxford University press, Delhi, 1994.

[193] K. Singh and V. Ballabh. People's participation in forest management: experience of van panchayats in U. P. hills. *Wasteland News*, **VI**(1):5–14, 1991.

[194] S. Sinha. Spiral or vortex? A review of the chakriya vikas pranali in Palamau district, Bihar. Economic Development Associates, Gurgoan, Sept. 1994.

[195] S. P. Sinha. 'Chakriya vikas ki kahani, vikas pathikon ki jubani' (The story of cyclic development, the road towards development). Patamu printing works, Daltenganj, 1996.

[196] E. Somanathan. Deforestation, property rights and incentives in central Himalaya. *Economic and Political Weekly*, January 26:PE37–PE46, 1991.

[197] SPWD. Joint Forest Management update 1993. Society for Promotion of Wastelands Development, New Delhi, 1993.

[198] D. O. Stahl II. The graph of prisoner's dilemma supergame payoffs as a function of the discount factor. *Games and Economic behaviour*, **3**(3):368–384, 1991.

[199] G. G. Stevenson. *Common Property Economics: a General Theory and Land Use Applications*. Cambridge University press, New York, 1991.

[200] M. Taylor. *The Possibility of Cooperation*. Cambridge University press, Cambridge, 1987.

[201] G. D. Thompson and P. N. Wilson. Common property as an institutional response to environmental variability. *Contemporary Economic Policy*, **12**(3):10–21, 1994.

[202] J. Tirole. *The Theory of Industrial Organization*. The MIT Press, Cambridge, Mass., 1989.

[203] A. Tiwari. Chakriya Vikas reverses educational system. Society for Hill Resource Management School, Daltenganj, 1992.

[204] UNEP. *Poverty and the Environment, Reconciling Short-term Needs with Long-term Sustainability Goals*. United Nations Environment Programme, Nairobi, Kenya, 1995.

[205] UP. Uttarakhand: ek dristi men. Vartaman Uttarkhand, 1995. Uttar Pradesh Government. A district-wise overview of major indicators in Uttarakhand (in hindi).

[206] N. Uphoff. Grassroots and NGO's in rural development: opportunities with diminishing states and expanding markets. *World Development*, **21**(4):607–622, 1993.

[207] E. van Damme. *Stability and Perfection of Nash Equilibria*. Springer-Verlag, Berlin, second revised and enlarged edition, 1991.

[208] V. Varalakshmi. Economics of bamboo basket making: a case study. Tata Energy Research Institute, 1992. *Joint Forest Management Series* **3**.

[209] V. Varalakshmi. Economics of goat and buffalo rearing - a case study. Tata Energy Research Institute, 1993. Joint Forest Management Series 4.

[210] V. Varalakshmi, R. Vijh, and S. S. Arora. Systems of *bhabbar* grass lease management in Haryana. Tata Energy Research Institute, 1993. *Joint Forest Management Series* **7**.

[211] R. Vijh and S. S. Arora. Joint forest management programme in Haryana: TERI's innovations. *Wasteland News*, **IX**(1):17–20, 1994.

[212] R. Wade. The management of common property resources: collective action as an alternative to privatisation or state regulation. *Cambridge Journal of Economics*, **11**:95–106, 1987.

[213] R. Wade. *Village Republics: Economic Conditions for Collective Action in South India*. Cambridge University press, Cambridge, 1988.

[214] J. M. Walker, R. Gardner, and E. Ostrom. Rent dissipation in a limited-access common-pool resource: experimental evidence. *Journal of Environmental Economics and Management*, **19**:203–211, 1990.

[215] J. M. Walker, R. Gardner, and E. Ostrom. Rent dissipation and balanced deviation disequilibrium in common pool resources: Experimental evidence. In R. Selten, editor, *Game Equilibrium Models, Vol. II, Methods, Morals and Markets*, pages 337–367. Springer verlag, Berlin, 1991.

[216] O. E. Williamson. *The Economic Institutions of Capitalism*. Free press, New York, 1985.

[217] O. E. Williamson. Reflection on the new institutional economics. *Journal of Institutional and Theoretical Economics*, **141**(1):187–195, 1985.

[218] O. E. Williamson. The new institutional economics: taking stock, looking ahead. *Journal of Economic Literature*, **XXXVIII**:595–613, September 2000.

[219] M. Woolcock. Social capital and economic development: toward a theoretical synthesis and policy framework. *Theory and Society*, **27**:151–208, 1998.

[220] H. P. Young. *Individual Strategy and Social Structure. An Evolutionary Theory of Institutions*. Princeton University Press, Princeton, New Jersey, 1998.

Index